| DATE | | | |
|---|---|---|---|
| | | | |
| | | | |
| | | | |
| | | | |
| | | | |
| | | | |
| | | | |
| | | | |
| | | | |
| | | | |
| | | | |
| | | | |
| | | | |

© THE BAKER & TAYLOR CO.

# ADAPTATION AND ENVIRONMENT

ROBERT N. BRANDON

# Adaptation and Environment

PRINCETON UNIVERSITY PRESS
PRINCETON, NEW JERSEY

*Library of Congress Cataloging-in-Publication Data*

Brandon, Robert N.
Adaptation and environment / Robert N. Brandon.
p. cm.
Includes bibliographical references.
ISBN 0-691-08548-X
1. Adaptation (Biology) 2. Natural selection. I. Title.
QH546.B73 1990
575—dc20      89-24206

# CONTENTS

In this book I explore a tightly interconnected set of questions within evolutionary biology whose common thead is the phenomenon of adaptation. I emphasize that the process of adaptation, and the products of that process, cannot be fully understood without an analysis of the notion of environment as it is relevant to the theory of evolution by natural selection. That theory is our only general and scientifically legitimate theory of adaptation. I argue that natural selection is the process of differential reproduction that is due to differential adaptedness to a common selective environment. Although this characterization of natural selection is relatively uncontroversial, I believe it is not well understood. I present the characterization in terms of two basic concepts: relative adaptedness (what many call "relative fitness"), and common selective environments. The former has received extensive attention from both philosophers and biologists interested in evolution, but the latter has been virtually ignored. When properly understood, the concept of common selective environments and the characterization of natural selection given in terms of it have surprising implications.

In chapters 1 and 2 I define the basic concepts. Chapter 1 is an overview of the concept of relative adaptedness, including the process of adaptation and the products of that process. I defend a version of what has come to be known as the propensity interpretation of adaptedness (or of fitness), while also reviewing the major rival interpretations, exposing the weaknesses of each. In chapter 2 I propose that in biology (or more accurately, within population biology) one can distinguish at least three different notions of environment: the *external environment*, the *ecological environment*, and the *selective environment*. There are interesting interrelations among these three sorts of environments, the most prominent among them being that the selective environment is decoupled from the external environment in the sense that variation in the external environment is neither a necessary nor a sufficient condition for variation in selective environments. I argue that it is the selective environment that is relevant to the theory of evolution by natural selection. This assertion has direct implications for both the theoretical and experimental study of some of the major questions of evolutionary biology, that is, those dealing with selection in heterogeneous environments. For instance, sexual reproduction and phenotypic plasticity, both pervasive biological phenomena, are thought to be selectively advantageous in heterogeneous environments.

In chapters 3, 4, and 5 I explore some of the ramifications of this conception of natural selection. Chapter 3 deals with questions concerning the level or levels at which selection occurs. I adopt David Hull's distinction of *interactors* and *replicators* (a generalization of the phenotype-genotype distinction). Selection (called *phenotypic selection* by quantitative geneticists) is an ecological process; as such it always occurs among interactors. For selection to result in evolution (i.e., transgenerational change), the variation on which selection acts must be heritable. This involves the process of replication. I argue that there is a hierarchy of plausible interactors, that is, a hierarchy of levels of selection, but that the replicators corresponding to each level of interaction do not form a neat hierarchy. This analysis differs significantly from those coming out of a population-genetics tradition, which conflate questions of levels of selection with questions of replication.

Chapter 4 is devoted to certain metatheoretical questions that concern the structure of the theory of evolution by natural selection. Although this is a philosophical exercise, the questions have been debated by reflective evolutionists for some time. For instance, what is the Principle of Natural Selection? Does it have empirical content? What is its role in the theory of evolution? Answers to these questions are directly relevant to how we distinguish selection from random drift, how we distinguish various levels of selection, and even how we distinguish the results of selection from those of random distribution of competing types across heterogeneous environments.

In chapter 5 I offer an account of what I call *ideally complete adaptation explanations*. According to this view, adaptation explanations are thoroughly mechanistic. This ideal is rarely, if ever, realized in practice because of the severe epistemological problems one encounters when trying to gather the requisite information.

But these problems do not detract from the overwhelming value of mechanistic explanations of how adaptations *could* evolve. Much recent criticism of the theory of natural selection, or of what Stephen Jay Gould and Richard Lewontin call the *adaptationist programme*, is based on the assumption that the *raison d'être* of the theory of adaptation is the explanation of particular traits. I argue that this is a mistaken assumption; that, although the explanation of particular adaptations is of interest in historical evolutionary biology, there is a purely process-oriented aspect of the theory to which such explanations are only tangentially relevant. Are adaptation explanations teleological? There is a sense in which they do explain teleological phenomena and in which they do answer a type of teleological question; but the explanations themselves are thoroughly mechanistic.

PHILOSOPHY of biology is an exciting field. And it surely is an interdisciplinary field. But dangers are inherent in such interdisciplinary work: one runs the risk of presenting baby philosophy for biologists and baby biology for philosophers. I hope I have been able to avoid that. I have tried to write this book not for two separate audiences—philosophers interested in biology and biologists interested in the conceptual foundations of their discipline—but for a single audience, one unified by a deep interest in evolutionary biology, especially in the theory of natural selection. The readers I have kept in mind are professional biologists or students of biology, and philosophers. I hope there will be some from other disciplines as well. I assume that the reader has a basic knowledge of genetics and statistics; those who lack it can consult one of the many textbooks that introduce this material far more effectively than I could. On the other hand, I have not assumed that the reader is thoroughly familiar with philosophical theories of causation and explanation. I am a philosopher, but I have studied evolutionary biology for the past fourteen years, and I do not try to classify problems as either philosophical or biological. Indeed, I doubt whether this could be done in a consistent and nonarbitrary manner and it is surely a fruitless exercise. Thus I have not tried to balance the "biological" and "philosophical" in order to satisfy a dual audience.

Although most biologists would readily admit to the importance of conceptual issues in evolutionary theory, many may doubt whether these issues have any direct impact on the practical concerns of evolutionary biologists. One of the purposes of this book is to show the tight relationship between conceptual issues and practical methodological issues. For example, the concept of selective environment that is introduced in chapter 2 has direct implications concerning both the importance and the design of "common garden experiments." In particular, when selection is frequency dependent, the proper design of a common garden experiment is different than when selection is not frequency dependent. This outcome runs contrary to some of our intuitions and expectations and has direct implications for the concept of evolutionary altruism (as discussed in chapter 3). Here, then, a conceptual analysis has ramifications for experimental design, and getting clear on proper experimental design helps resolve another conceptual question.

My goal is to answer the questions I pose and to convince the reader of the correctness of my solutions. I realize that I may fall short of that goal. But if I succeed in communicating the importance of some topics, if I succeed in showing the interrelations of topics that are not normally thought to be related, then my efforts will have been worthwhile.

*Durham, N.C.*
*September 1989*

Part of the work on this book was supported by NSF Grant #SES86-18442. Some empirical work related to chapter 2 was supported by the Duke University Research Council. Without this support my book would have been delayed or, perhaps, never written.

I know I could not have written this book without the help of many friends and colleagues. The following people read either part or all of my manuscript: John Beatty, Robert Boyd, Richard Burian, Henry Byerly, Kurt Fristrup, Michael Ghiselin, Marjorie Grene, Henry Horn, David Hull, Egbert Leigh, Marcia Lind, Rick Michod, Mark Rausher, Robert Richardson, and George Williams. Their comments have been invaluable. I owe a special debt to two of my colleagues at Duke, Janis Antonovics and Brent Mishler. Besides having carefully read each chapter, they were a constant source of encouragement and advice. In particular, chapter 2 is very much the product of my interaction with Antonovics. If the book has too much of a botanical ring to it, it is because I hung out too much with the damned botanists.

I owe a special debt of a different sort to my wife, Gloria Meares. She provided needed support during the difficult times I experienced in writing this book; she also carefully read the manuscript. On the other hand, my daughter Katherine was of no help whatsoever. She should have had the good sense not to have been born while I was trying to write a book.

At this point it is customary for an author to absolve those whose contributions have just been acknowledged from any blame for errors that remain. In the spirit of the Reagan '80s, I want to go one step further: I will frankly admit that mistakes were made; but these are free-floating mistakes for which no one should be blamed.

ADAPTATION AND ENVIRONMENT

# Adaptation and Natural Selection

The existence of adaptations, the fit between organisms and their environments, is one of the most striking features of the biological world. Before Darwin (1859) numerous accounts were offered to explain adaptations, the most prominent among them being the creationist account. According to this account, organisms were designed by God to fit the demands of their environments. Darwin offered an alternative account: the theory of evolution by natural selection. There have been still other rival explanations of adaptations. Perhaps the most important of these is a "Lamarckian" theory whereby organisms somehow adapt to their environment during their lifetime and then pass on these acquired adaptive characteristics to their offspring. As is well known, Darwin put some credence in this account. Indeed, we should too, but only for those organisms capable of behaviorally transmitting information across generations. For example, in a well-studied colony of Japanese macaques, a single individual learned to separate wheat from its chaff by throwing it into water. This adaptive behavior was transmitted to other members of the colony and passed down to later generations so that this once absent behavior is now quite common. Although this is a mechanism by which organisms can adapt to their environments, it is probably quite limited and explains only a small fraction of the adaptations in nature.[1] As the creationist and "Lamarckian" theories illustrate, the theory of evolution by natural selection is not the only possible theory of adaptation, but for now it is by far the best theory we have to explain the bulk of adaptations in nature.

In this chapter we will explore three related concepts of adaptation: the concept of relative adaptedness, the concept of the process of adaptation, and the concept of adaptation as the product

[1] See Bonner (1980), Boyd and Richerson (1985), and Brandon (1985b) for further discussion.

3

of that process. We will examine their interrelations and the roles they play in the theory of evolution by natural selection.

## 1.1. A SIMPLE CASE OF EVOLUTION BY NATURAL SELECTION

Let us start by describing a simple case of evolution by natural selection. Suppose that in a given population of organisms there is directional selection for increased height. To say that there is directional selection for increased height is to say that taller organisms have (or tend to have) greater reproductive success than shorter ones, that is, that reproductive success is an increasing function of height. The ecological reasons for this can be indefinitely varied. For instance, in one ecological setting taller plants may receive more sunlight and so have more energy available for seed production. In another setting, taller animals may be more resistant to predation. Also, differences in reproductive success can result from differences in fecundity (as in the first case) or from differences in survivorship (as in the second case) or from still other causes (e.g., differences in mating ability). To fix ideas, let us suppose that in our case taller organisms have a higher survivorship. Thus we can suppose that prior to selection the distribution of height in the population is as shown in figure 1.1. Selection occurs by differential survivorship. The postselection distribution

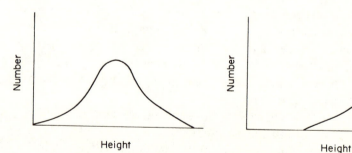

Figure 1.1. Schematic illustration of preselection distribution of height in a population.

Figure 1.2. Schematic illustration of postselection distribution of height in a population.

of height is given in figure 1.2. This is the first step in the process of evolution by natural selection.[2]

From the point of view of population biology, evolution is any change in the distribution of "types" over *generational* time. Population geneticists define evolution as any change in the relative frequency of alleles over generational time. If a more organismic approach is preferred, evolution could be defined as any change in phenotypic distributions over generational time. The important point is that evolution involves change over generational time. Thus in our example, although selection has produced change, evolution has yet to occur.

The next step in this process involves reproduction. We can assume that all organisms surviving to this stage will have equal reproductive success. But because of viability selection a disproportionate number of taller organisms is left to reproduce. Through reproduction the next generation is formed. If selection is to result in evolutionary change in this case, height must be heritable. Phenomenologically, that means that taller than average organisms must tend to have taller than average offspring, and shorter than average organisms must tend to have shorter than average offspring. Of course, in standard cases genes, not height, are directly transmitted from parent to offspring. Thus taller parents tend to have genotypes different from those of shorter parents, and these genes are transmitted to the respective offspring. The offspring of taller parents tend to have genotypes that are different from those of offspring of shorter parents. According to the population-genetic definition of evolution, evolution has occurred.

But in order to go full circle, in order to get to the stage where selection occurs in the offspring generation, a final step is required. These differing offspring genotypes must develop. During the process of epigenesis (i.e., the process of the development of the fertilized egg into a mature organism), these genotypic differences are translated into phenotypic differences. Thus, in our case, the offspring distribution of height is shifted to the right of that of the preselection distribution of the parental generation (as shown in figure 1.1). How far it is shifted depends on the degree of heri-

---

[2] Mayr (1978) has described evolution by natural selection as a two-step process. My description essentially follows his and will be discussed in greater detail in chapter 3.

tability (and the strength of selection, which is represented by the difference between figures 1.1 and 1.2). In any case, evolution has occurred; the frequency distribution of height has changed over generational time.[3]

## 1. 2. DARWIN'S THREE PRINCIPLES

The case described above is a simple example of evolution by natural selection.[4] By examination we can abstract from it the conditions necessary for the evolutionary process to work.

The first condition is that there be variation in height. To say that there is directional selection for increased height, to say that taller organisms have greater reproductive success than shorter organisms, implies that there is variation in height. The selection process we described worked by means of differential survival; but whatever the means of selection, there can be no selection for increased height if there is no variation in height. Height is a morphological trait. We can also imagine selection on physiological or behavioral traits, or more generally on any sort of *phenotypic* trait. Thus the first condition is that there must be some variation in some phenotypic traits.[5]

For selection to occur there must be variation, but a second condition is necessary for selection to have an evolutionary effect. Recall that in the above example selection changed the population distribution of height from that of figure 1.1 to that of figure 1.2. But this is not evolutionary (i.e., cross-generational) change; for

---

[3] I have distinguished three steps in the process of evolution by natural selection: selection, the differential replication of genes, and development. A discussion of why this process is described as a three-step rather than a two-step process can be found in chapter 3.

[4] Of course, no one example is fully representative of all cases. My example is special in two regards. First, it is a case of directional selection, which favors one extreme in a phenotypic distribution. Other types of selection are stabilizing, which favors the mean phenotype, and disruptive, which favors both ends of the distribution (see Endler 1986, pp. 16ff.). Second, in my example selection occurs by means of differential survivorship rather than by means of differential fecundity, differential mating ability, or other possible means. The special features of the example do not affect the conclusions we will draw from it.

[5] I am assuming here that, to use Mayr's words, "natural selection favors (or discriminates) phenotypes, not genes or genotypes" (1963, p. 184). I will argue for this position in chapter 3.

such a change to occur, height must be heritable. Presumably in our case the mechanism of inheritance is the transmission of genetic differences. But for present purposes the mechanism is irrelevant; what matters is that taller than average parents produce taller than average offspring. Thus the second condition necessary for evolution by natural selection is that the relevant phenotypic traits be heritable.

A third requirement—not surprisingly—is that there be selection. In our example selection took the form of differential survival. But no matter what form selection takes—differential survival, differential fecundity, or differential mating success—the result is the same, that is, there is differential reproductive success. In our case taller organisms have, on average, greater reproductive success than shorter ones. Differential reproductive success, often termed differential fitness, is the third condition necessary for evolution by natural selection.

In summary, then, we have the following three conditions:

1. *Variation*. There is variation in phenotypic traits among members of a population.
2. *Inheritance*. These traits are heritable to some degree, meaning that offspring are more like their parents with respect to these traits than they are like the population mean.
3. *Differential reproductive success*. Different variants leave different numbers of offspring in succeeding generations.[6]

I have argued that these three conditions are *necessary* for evolution by natural selection. But are they *sufficient* conditions as well? This is a twofold question. First, are the conditions sufficient to produce evolutionary change, that is, will evolution occur when these conditions are met? Second, are the conditions sufficient to produce evolution *by natural selection*?

Contrary to Lewontin (1968, 1970, and 1978), Wimsatt (1981, 1982), and others (including Brandon 1978), these three conditions are not jointly sufficient to produce evolutionary change. The conditions may be satisfied and yet the population can remain at equilibrium when inheritance is biased in a way that exactly counter-

---

[6] These three conditions have been formulated by Lewontin (1968, 1970, and 1978). The wording here is my own but is closest to Lewontin (1978).

balances selection. Although this fact is interesting, it is not relevant to our present concerns. What is relevant is that, even ignoring the complications of biased inheritance, the three conditions are not jointly sufficient to produce evolution *by natural selection*. A closer examination of the three conditions will verify this statement.

We can certainly ask why any of these conditions obtain. Why is there variation in a population? Why are some traits heritable? Why do some organisms outreproduce others? Much work in twentieth-century population biology has been devoted to exploring the first question. Many questions still remain unanswered, but we have learned much about the origins and maintenance of variation in populations (Lewontin 1974). Likewise, twentieth-century genetics and population genetics have shed considerable light on the inheritance of traits in populations. But what of the third question? What is the explanation for differential reproductive success?

We must consider two possible explanations. One is that differential reproduction is the result of natural selection, but let us first consider an alternative explanation. In a finite population, especially in a relatively small population, some types of organisms may outreproduce others simply by chance. In other words, by chance alone one can get a positive correlation between trait value (e.g., height) and level of reproductive success. Imagine a population where just prior to reproduction every organism flips a (fair) coin. If the coin lands on heads the organism reproduces, if it lands on tails the organism dies. In a relatively small population it is not unlikely that a disproportionate number of taller organisms get lucky. Thus condition (3) will be satisfied, and provided that condition (2) is also satisfied, the next generation's trait distribution will be altered, that is, evolution will occur.[7] What will happen if this process continues for many generations? For simplicity, suppose that there are only two types, tall and short, and that they accurately reproduce their type. Then, no matter what the initial ratio of tall to short, we can be sure that one type will eventually go to fixation in the population by a process statisticians call *ran-*

[7] Note that condition (3), differential reproductive success, implies condition (1), variation.

*dom walk.* This is evolution, but not evolution by natural selection; thus conditions (1)–(3) do not assure that evolution by natural selection will occur.

It is important to note that random walk—or, to use the term more common in evolutionary biology, *random drift*—is an inevitable consequence of finite population size; therefore, as a process, it is not a mutually exclusive alternative to the process of natural selection. That is, when natural selection occurs in a real (and therefore finite) population, random drift will occur as well (see Beatty 1984). The effectiveness of drift depends on the size of the population and the strength of selection (see Hartl 1988, pp. 176–177). But as explanations of differential reproduction, drift and natural selection are alternatives. Thus in offering an alternative explanation of (3), the natural-selection explanation, we must keep in mind the possibility of explaining (3) by means of random drift. Natural-selectionist explanations should be testably different from explanations based on chance.

## 1.3. THE PRINCIPLE OF NATURAL SELECTION

Darwin's contributions to evolutionary biology are so numerous that singling out any one of them as *the* most significant is bound to be controversial. But it is safe to say that his theory of natural selection is one of his major contributions.[8] This theory is designed to explain condition (3), differential reproductive success, and I believe it ultimately explains the origins and maintenance of adaptations in nature.

What is the Darwinian explanation of (3)? The standard story is that Darwin explained (3) by his postulate of the "struggle for existence" (or in Spencer's words, which Darwin later used, "the survival of the fittest"). How does this explain (3)? To answer this question, we must first define some terms.

Condition (3) is about differential reproductive success. Reproductive success is sometimes called *Darwinian fitness* or sometimes just *fitness*. It refers to the actual genetic contribution to succeeding

[8] This major contribution had little effect on Darwin's contemporaries. He did convince the majority of the scientific community that evolution has occurred and does occur, but very few accepted his natural-selection explanation of it. See Hull (1973).

generations. (Whether "succeeding" means the next generation, the next few generations, or generations in the distant future will be discussed shortly.) This contribution can usually be measured in terms of the actual number of offspring left by the target organism, or in terms of the mean of the actual number of offspring left by members of the similarity class to which the target organism belongs (e.g., all organisms with genotype $G_1$). The terminology used here is confusing. Although the terms "Darwinian fitness" and "fitness" are often used in just the way described above, they are sometimes used to refer to the potential, rather than the actual, genetic contribution to the next generation. This distinction is very important. It is unfortunate that the same terms have been used to name two closely related and easily confused concepts. One way of marking the distinction is to use the terms "actualized fitness" or "realized fitness" to refer to actual reproductive success, and "expected fitness" to refer to the potential reproductive success. For reasons I will discuss later, I prefer the term "adaptedness" to "expected fitness" but will use both. But to avoid confusion in this chapter I will rarely use "fitness" unmodified; thus I will talk about realized fitness and expected fitness but not simply fitness, unless the context does not demand the distinction (as is the case in theoretical models where there is usually no distinction, and no need to make the distinction, between realized fitness and expected fitness).

Now to return to our question: How does "the struggle for existence" or "the survival of the fittest" explain (3)? Condition (3) implies that actualized fitness is correlated with certain phenotypic traits. Why does this correlation exist? Why is there differential actualized fitness? Darwin's answer, which he arrived at after reading Malthus's *Essay on Population* (1798), was the following: since in each generation more individuals are produced than can survive to reproduce, there is a struggle for existence. In this "struggle" (which in its broadest sense is a struggle of the organism with its environment, not just with other individuals; see Darwin 1859, p. 62) certain traits will render an organism *better adapted* to its environment than conspecifics with certain other traits. The better adapted individuals will tend to have greater reproductive success than the less well adapted. Why do some organisms have greater reproductive success than others? The Darwinian answer is this: they are (for the most part) better adapted to their environment.

What does this explanation presuppose? It seems to presuppose the following as an explanatory principle or as a law of nature:

PNS: If $a$ is better adapted than $b$ in environment $E$, then (probably) $a$ will have greater reproductive success than $b$ in $E$.

I will argue that this principle or law, which I will call the *principle of natural selection* (PNS), provides the basis of Darwinian, or natural-selectionist, explanations of differential reproduction. That is, natural-selectionist explanations explain differential reproductive success in terms of differential adaptedness to a common environment. The exact role and the logical status of this principle need considerable clarification and are the focus of chapter 4. But clearly everything depends on how the concept of relative adaptedness, the concept of one organism being better adapted than another to their common environment, is defined.

## 1.4. Three Approaches to Defining Relative Adaptedness

There are three fundamentally different approaches to defining relative adaptedness.[9] First, we might define it in terms of actual reproductive success.[10] This definition has the advantage of being transparently operational, but has the disadvantage of being explanatorily empty. If we define it in terms of relative reproductive success, then we would be asking the phenomenon of differential reproductive success to explain itself. Clearly, if we are to have an explanatory PNS, this won't do. Natural selection is not just dif-

[9] I discuss these approaches in much greater detail in Brandon (1978). Mary B. Williams (1970; also see Rosenberg 1982, 1983) argues that adaptedness, or fitness, should not be defined. This avoids the questions with which we are presently concerned and so is of no relevance. See Brandon and Beatty (1984) for further discussion.

[10] See Stern (1970, p. 47) where he quotes Simpson, Waddington, Lerner, and Mayr to this effect. (Mayr, I think, was quoted out of context; see Mayr 1963, pp. 182–184.) Stern approves of this definition. Also see Beatty (1984, p. 191) for further discussion and references.

ferential reproduction, but rather it is the differential reproduction that is *due* to differential adaptedness, that is, due to the adaptive superiority of those who leave more offspring. Alternative ways of saying the same thing are: (1) natural selection is nonrandom differential reproduction; and (2) natural selection is differential reproduction that is in accord with expected reproductive success. (The meaning of "expected" relevant here will be discussed shortly.) Richard Burian (1983, pp. 302–303) argues that this basic understanding of natural selection is implicit in the work of Dobzhansky and Simpson. I suggest that it is implicit throughout most of contemporary evolutionary biology, a fact that is obvious when one reflects on what constitutes an adequate experimental design for detecting natural selection either in nature or in the lab. (See Endler 1986 for a useful summary of the methods used for detecting natural selection.) If we defined relative adaptedness in terms of actual reproductive success, we would not have to worry about repeatability or sophisticated statistical designs in doing experiments to show that one type of organism is better adapted than another, or to show that natural selection is occurring. But experimental evolutionary biologists do worry about such things.

The point of all of this can be illustrated by a simple story. Suppose we place two impregnated dogs, one a basset hound and the other a German shepherd, on an island to see which is better adapted to its ecological conditions. Preliminary ecological analysis may lead us to expect that the shepherd would be much better adapted than the basset because of the types of prey available. But lightning strikes and kills the shepherd, and the basset survives and manages to raise her litter. Obviously we do not conclude from this that the basset is better adapted to the island environment. As mentioned above, that is why in designing experiments to detect selection we use multiple replicates. The point of this story is that chance can disassociate adaptedness from actualized fitness, an event especially likely to occur in small populations (or when selection is particularly weak). Thus if we want to distinguish natural selection from drift, if we want natural-selectionist explanations of evolutionary change to be testably different from explanations in terms of random drift, then we must adopt some concept of relative adaptedness that is not tied directly to actual reproductive success.

Second, we might define relative adaptedness in terms of some specific biological property, such as height, strength, fecundity, or energetic efficiency.[11] Philosophically, this is the most attractive option. If we could find some biological property, whether a straightforward one like height or a more complex one like energetic efficiency, that was selectively advantageous in all environments for all organisms, then we could use variation in that property to *explain* differential reproductive success. Unfortunately, nature is much too variegated to make this approach workable. For instance, although fecundity might be an obvious candidate for a property that is always selected for, there is often no selection for higher fecundity.[12] Similarly for any other specific biological property. Selective environments can differ radically from one another, and the ways in which organisms can be better adapted than their conspecifics to those environments can differ radically as well.[13] This point has been widely accepted by philosophers of biology and has been labeled the *supervenience* of adaptedness (or of fitness; see Brandon 1978, Rosenberg 1978, Mills and Beatty 1979, Sober 1984, and Rosenberg 1985). To say that adaptedness supervenes on basic phenotypic properties of organisms (in environments) implies two things: (1) The adaptedness of an organism in an environment is determined by its phenotypic features so that any two organisms in the same environment with the same phenotypic features have the same adaptedness; and (2) there is no manageable set of these phenotypic properties in terms of which we could define adaptedness. In other words, we cannot define

[11] Bock and von Wahlert (1965) try to do this in terms of energetic efficiency. In Brandon (1978) I argue at length against their approach.

[12] As shown by Lack (1954). This must be quite surprising to those with only a superficial understanding of the theory of natural selection. For example, Popper (1972, p. 271) thinks it is "one of the countless difficulties of Darwin's theory" that natural selection should do anything other than increase fecundity. The explanation is quite simple: increased fecundity often results in a decreased number of offspring surviving in the next generation. See Williams (1966, chap. 6) for discussion.

[13] In Brandon (1978) I give two related arguments against the possibility that one could adequately define relative adaptedness in terms of some specific biological property. The first argument is that unmanipulated nature is too varied. The second is that we human beings are a part of nature, and so we can design an artificial selection experiment to falsify anyone's claim that biological property X is always selected for. Since we are a part of nature, artificial selection is a type of natural selection.

relative adaptedness in the following way: *a* is better adapted than *b* in *E* if and only if *a* is taller than *b* or *a* is more fecund than *b* or *a* is more energetically efficient than *b* or. . . . The problem is in filling in the dots.

The third basic approach to defining relative adaptedness is what I will call the *propensity interpretation of adaptedness*. According to this approach, relative adaptedness is defined in terms of the abilities or capacities of the organisms to survive and reproduce in their environment. Unlike the first approach, these abilities or capacities are not identified with actual reproductive success. Unlike the second approach, the ability to survive and reproduce is not a *specific* biological property since it will be differently instantiated for different organisms and different environments.

## 1.5. THE PROPENSITY INTERPRETATION OF ADAPTEDNESS

In chapter 4 we will explore the role of the PNS in the theory of evolution by natural selection, and I will argue that it is the central explanatory principle of that theory. The aim of the propensity interpretation of adaptedness is to define relative adaptedness in a way that will allow the PNS to play that role. To do this, we would like to define relative adaptedness in a way that renders the PNS (a) true, or empirically correct; (b) epistemologically applicable and/or testable; and (c) general. These desiderata probably do not need much defense, and I think they will be fully understood when we develop an adequate definition of relative adaptedness. So for the moment I will only offer a brief justification of each. False statements, or statements that do not fit the empirical facts, explain nothing.[14] Thus (a). Furthermore, if the PNS is to be explanatory we need to know how to apply it to the cases we wish to explain. The flip side of this is that application of the PNS to particular cases should allow us to test it, that is, to see whether it truly describes the case. Thus (b). Finally, we would like to have a PNS that is general, that is, that applies to all organisms under all selective circumstances. Thus (c).

[14] Philosophers would distinguish truth from empirical correctness. Empirical correctness is a necessary condition for truth, but not a sufficient one. For our purposes this distinction is not important.

It should be clear that we can explain the actual behavior of an object in terms of its abilities, capacities, or propensities. For instance, one can make a statement about the top speed of a car that is independent of the top speed the car has actually achieved. Indeed, this is an instructive example, since there already are methods for estimating top speed that do not require one actually to drive a car. One can calculate top speed using various criteria—the car's weight, horsepower, top gear ratio, tire circumference, and aerodynamic drag. This calculated capacity or ability is potentially explanatory—for instance, it could be used to explain why one car outran another in a long drag race. Likewise, we could well explain why one organism outreproduces another in a common environment in terms of their differing abilities or capacities. This possibility suggests the following definition of relative adaptedness:

(RA): *a* is better adapted than *b* in *E* iff *a* is better able to survive and reproduce in *E* than is *b*.

("iff" is short for "if and only if.") This definition avoids tautology, that is, it is not given in terms of actual reproductive values. It is also a general definition, holding for all organisms in all environments. But how can we apply it to particular situations? It is clear that as it stands, (RA) is not epistemologically applicable, that is, we do not know how to apply it to particular cases. The reason for this is straightforward: we have no theory nor any data base that tells us, for any arbitrarily chosen group of organisms in some environment, what allows one organism to be better able to survive and reproduce than another organism in the same environment.

What is it for one organism to be better able than another to survive and reproduce in an environment? For particular organisms in particular ecological settings we might be able to answer this question; but is there a general answer as well? In the sense that would imply some specific biological property that is inevitably selected for, I have already argued that there is no general answer. But in another sense I believe that there is. If the PNS is to be explanatory, we cannot explicate the ability to survive and reproduce in terms of that which it is supposed to explain, namely, actual reproductive success. But suppose we explicate the ability in terms of the *probability* to have a certain level of reproductive success. This will work only if we have a way of conceptualizing the probable that does not equate the probable with the actual in some finite population. (Recall the need to distinguish natural selection from random drift. If we equate the probable with the actual in some finite population, then under the propensity interpretation of adaptedness we would not be able to differentiate natural selection from random drift.)

Two interpretations of probability meet this need. One is called the *limit relative frequency interpretation*, according to which the probability of an event (e.g., a coin landing on heads) is the limit of the relative frequency of that event in an infinite series of trials (see Reichenbach 1949). The observed relative frequency of the event in some finite sequence is evidence for a certain probability judgment, but is not definitive of that probability. (We might call the interpretation of probability that equates the probable with the actual in some finite sequence the *finite relative frequency interpretation*. This is an interpretation that has had very few adherents, but see Russell 1948 for one.) The other relevant interpretation is called the *propensity interpretation* (Popper 1959, Hacking 1965). According to this interpretation the probability of a coin landing on heads is a property of that coin (or of the coin and tossing device) just as the solubility of a sugar cube is a property of the sugar cube. These properties are dispositional, that is, they manifest themselves only under certain conditions. The sugar cube manifests its water solubility when in water. The coin manifests its probability of heads when tossed. According to standard views of dispositional properties (e.g., Quine 1960), the sugar cube's solubility is based on some more basic nondispositional physical properties—in this case, on the sugar's molecular structure. Likewise for the coin.

Epistemologically, the propensity interpretation allows for two quite different avenues to estimates of probability. One is via observed relative frequencies, the other by means of the physical properties that underlie the propensity. That is, we can flip the coin a sufficient number of times, or we can investigate the physical properties (such as its shape, its internal structure, the surrounding physical forces, etc.) relevant to its behavior when tossed. This is analogous to our epistemological situation with respect to other dispositional properties, such as water solubility or top speed of a car. In the former case we can either drop an object in water or investigate its internal molecular bonds to determine whether or not it is water soluble. In the latter case we can drive the car at full throttle in top gear or we can calculate its top speed based on its weight, aerodynamics, tire circumference, gear ratio, and horsepower. Ontologically, however, the dispositional property cannot be identified with its manifestation, but rather it must be identified with its underlying physical basis. Thus to be water soluble just means to have a certain molecular structure (a structure that is in fact well understood). And the probability of getting heads on the toss of a coin just means the coin has the relevant set of physical properties for that occurrence. (These properties are probably not well understood.)

Although both interpretations of probability meet our needs, I prefer the propensity interpretation for two reasons. First, it is, at worst, epistemologically on a par with the limit relative frequency interpretation. Both interpretations can make free use of observed frequencies in estimating probability. But the propensity interpretation can make better sense of the use of the other sort of data (those concerning the underlying physical properties) in estimating probabilities than can the limit relative frequency interpretation. Second, from an explanatory point of view it is far superior. The limit relative frequency interpretation is based on fiction. There are no infinite sequences of tosses of coins, there are no infinite biological populations. On the other hand, the propensity interpretation identifies probability with those factors causally responsible for the observed frequencies.

Let us return to the problem of defining relative adaptedness. Suppose that from our ecological theories and/or from observed reproductive success we could estimate the distribution of probabilities of the number of offspring left by a given organism in a given environment. That is, for organism $O$ in environment $E$ we have a range of possible numbers of offspring, $Q_1^{OE}$, $Q_2^{OE}$, . . . , $Q_n^{OE}$, and for each $Q_i^{OE}$ our theory/observational data associate a number, $P(Q_i^{OE})$, which is the probability (or chance or propensity) of $O$ leaving $Q_i$ offspring in $E$. Given all this, we define the adaptedness of $O$ in $E$ (symbolized as $A(O,E)$ as follows:

$$A(O,E) = \Sigma \, P(Q_i{}^{OE})Q_i{}^{OE},$$

that is, the adaptedness of $O$ in $E$ equals the expected value of its genetic contribution to the next generation. (The units of value are arbitrary. All that matters here are the ordinal relations among the numbers associated with each pair $\langle O,E \rangle$.) Given this definition of adaptedness we can now define relative adaptedness as follows:

(RA') $a$ is better adapted than $b$ in $E$ iff $A(a,E) > A(b,E)$.

Two things should be clear. First, (RA') makes sense only for intra-specific, intraenvironmental comparisons. Second, (RA') is a step in the right direction only with a proper interpretation of probability. (I have mentioned two interpretations that can do the job and have argued for the superiority of the propensity interpretation.)

Recall that we want our definition of relative adaptedness to fit the facts of natural selection. In this way the concept of relative adaptedness is dependent on that of natural selection. Recall also that I have suggested that natural selection be defined in terms of relative adaptedness. This conceptual interdependence may seem problematic, but in chapter 4 I will argue that it is not. For now let us concentrate on the desideratum that our definition of relative adaptedness is empirically correct (i.e., that it fits the facts of natural selection). We can test the empirical correctness of a definition of relative adaptedness by comparing it to clear cases of natural selection, whether those cases are empirical or theoretical. (RA') is a step in the right direction, but there is an important theoretical reason why it is not empirically adequate.

The definition of adaptedness on which (RA') is based is the probabilistic analog of an arithmetic mean. Consider a population consisting of two types of organisms, $G_1$ and $G_2$. (Reproduction is asexual and true to type, i.e., $G_1$'s produce $G_1$'s and $G_2$'s produce $G_2$'s.) $G_1$'s always produce exactly two offspring, while $G_2$'s produce one offspring 50% of the time and three offspring the rest of the time. This variation may be temporal; for example, in one generation the $G_2$'s may all produce one offspring each, in the next generation they may all produce three offspring each. Or it may be spatial in that in some locations of the range of the population, $G_2$'s produce one offspring each while elsewhere in that range they produce three each. Finally, the variation may be developmental, that is, there is no spatial or temporal correlation with off-spring number, but still 50% of $G_2$'s produce one offspring and the rest produce three. Clearly the two types have the same arithmetic mean actualized fitness, and, according to the above definition, they have the same adaptedness. But John Gillespie (1973, 1974, and 1977) has shown that the two types do not have the same evolutionary fate. Natural selection favors $G_1$ over $G_2$.

The difference between $G_1$ and $G_2$ is not in the mean number of offspring, but rather in the variance in the number of offspring. Gillespie (1977) has shown that this increased variance is always selectively disadvantageous.[15] We can use the above example to illustrate this point. Suppose that $G_2$'s variation is temporal, that in one generation $G_2$'s all have one offspring, and in the next generation they all have three. Thus in some generations they will have 50% more offspring on average than the $G_1$'s, in the other generations they will have 50% fewer. Suppose that in the $t$th generation there are equal numbers of $G_1$'s and $G_2$'s. Let $N$ be this number. If $t$ is a good generation for $G_2$'s, then there will be $3N$ $G_2$'s in the $(t + 1)$st generation. If it is a bad generation for them, then there will be $N$ $G_2$'s in the $(t + 1)$st generation. In either case there will be $2N$ $G_1$'s in the $(t + 1)$st generation. In the first scenario the relative frequency of $G_2$'s to $G_1$'s goes from .50 in the $t$th generation to .60 in the $(t + 1)$st. In the second scenario it goes from .50 in the $t$th generation to .33 in the $(t + 1)$st. Thus there is a 10% advantage to the $G_2$'s in good generations, but a 17% dis-advantage in bad generations. (These percentages obviously de-pend on the initial relative frequency, but the qualitative relation holds regardless of initial frequency.) Thus the advantage of hav-ing one offspring more than the mean is more than offset by the disadvantage of having one less than the mean. One can do a sim-ple simulation to see how this works over a number of genera-tions. In the $t$th generation there are $N$ each of $G_1$'s and $G_2$'s. Let $t$

---

[15] John Beatty and Susan Finsen (1989) have questioned this assertion based on an argument I find unconvincing. Ekbohm, Fagerstrom, and Agren (1979) also question Gillespie's conclusion. They point out that in many biologically realistic cases, reducing variances in offspring numbers will have certain associated costs. Thus in such cases there is no directional selection for minimal variances. I agree with their conclusions, but I think that they do not touch Gillespie's theoretical point, which is that *everything else being equal* increased variance is selectively dis-advantageous.

be a good generation for $G_2$'s. Thus there are $2N$ $G_1$'s and $3N$ $G_2$'s in the $(t + 1)$st generation. Alternate the good and bad generations for $G_2$'s. The $(t + 4)$th generation will be the last one where the $G_2$'s relative frequency exceeds .50 (it will be .53). By the $(t + 10)$th generation there will be $512N$ $G_1$'s and $243N$ $G_2$'s, so the relative frequency of $G_2$ will be .32. It will continue to decline.

Thus (RA') must be rejected because the definition of adaptedness on which it is based is not sensitive to the variance in offspring number (while natural selection is). Clearly we need to discount the mean value by some function of the variance. What function? Gillespie (1973, 1977) argues that when the variation is temporal (as in the example above) the mean value should be discounted by $1/2\sigma^2$, where $\sigma^2$ is the variance in offspring number. On the other hand, when the variation is developmental the mean should be discounted by $1/N\sigma^2$, where $N$ is the population size (Gillespie 1974, 1975, 1977). Let '$f(E, \sigma^2)$' denote some function of the variance in offspring number for a given type, $\sigma^2$, and of the pattern of variation. These patterns of variation in realized fitness will be discussed extensively in chapter 2. They are properties of the environment, $E$ (except for truly random developmental variation). I have just mentioned two forms this function can take, and surely there are others (see Lacey et al. 1983 for a discussion of the theory and applications to real populations). In any case, we can now define the adaptedness of an organism $O$ in environment $E$ as follows:

$$A^*(O,E) = \Sigma P(Q_i^{OE})Q_i^{OE} - f(E,\sigma^2).$$

This is the relevant expected value of $O$'s genetic contribution to the next generation in $E$.[16] With this definition in hand we can now define relative adaptedness as follows:

(RA''): $a$ is better adapted than $b$ in $E$ iff $A^*(a,E) > A^*(b,E)$.

[16] In standard mathematical usage "expected value" refers to the mean value, not the mean discounted by some function of the variance. In what follows, when I talk of expected reproductive success I will be using the word "expected" in the extended sense defined above. Beatty and Finsen (1989) show that the skew of a distribution of offspring numbers, as well as the mean and variance, also matters. That is why in the above definition of $A^*(O,E)$, the function $f(E,\sigma^2)$ is a dummy function in the sense that the form it takes can be specified only after the details of the selection scenario have been specified.

Recall that our informal definition of relative adaptedness, (RA), was general and potentially explanatory, but it was not epistemologically applicable. How does (RA″) fare on our desiderata? Given the propensity interpretation of probability it is potentially explanatory. In this interpretation the probability of reproductive success, or expected genetic contribution to the next generation, is a property of the organism in its environment (just as the probability of heads for a coin is a physical property of the coin and of the tossing device). The organism in its environment has this property even if it is struck by lightning prior to leaving any offspring (just as the chance of heads may be 1/2 for a coin even if it is unique and is melted before it is ever tossed). Thus (RA″) is independent of actual reproductive values. The occurrence of "probably" in the PNS may be confusing, but (RA″) does not turn the PNS into a tautology.[17] (RA″) clearly satisfies condition (c), that is, it is general. Like (RA), (RA″) is not empirically incorrect, and so we will say it satisfies (a), that is, it is empirically correct. But does it satisfy (b), that is, is it epistemologically applicable? To some extent it is. Unlike (RA), which hardly constrains what is to count for or against relative adaptedness, (RA″) offers substantial constraints. That it does so is clear from the above rejection of (RA′), which was applicable enough to show that it does not fit certain cases of natural selection. But still (RA″) does not satisfy (b). It would if we had a single, all-encompassing theory of adaptedness from which we could derive the adaptedness (as defined above) of any organism in any environment. But presently there is no such theory. Indeed, I have argued that no such theory is possible (see note 13).

The suggested definition is useful as what I will call a *schematic definition*. It is neither applicable nor testable, but particular *instances* of it are both. What do I mean by an instance of (RA″)? Formally, in an instance of (RA″) we fix the value of the environmental parameter $E$, and limit the range of the individual variables $a$ and $b$ to a particular population of organisms living in $E$.[18] Such

---

[17] The PNS becomes an instance of the Law of Likelihood or the Principle of Direct Inference. It is analogous to the following: If the chance of heads for coin $a$ is 1/2 and the chance of heads for coin $b$ is 1/4, then (probably) when both coins are tossed a small number of times $a$ will land on heads more than $b$. This point will be discussed in greater detail in chapter 4.

[18] Actually (RA″) is doubly schematic. It is schematic in the sense just specified,

an instantiation would represent a hypothesis concerning the relative adaptedness of the various types in the population. Two kinds of data are relevant to framing such hypotheses. The first comes from what we might call ecological/engineering analyses. Here we study the different variants—morphological, physiological, and behavioral—and how they interact with the environment. The second comes from detailed observations, usually in controlled experiments, of the performance of the variants within common environments. (The experiments will be discussed in chapter 2.) These two kinds of data are analogous to the two kinds of data relevant to hypotheses about dispositional properties, such as water solubility and top speed, as discussed earlier. Because adaptedness is a dispositional property, this similarity should not be surprising.

Let me give a simplified example using only the first kind of data. Suppose that in a certain population of moths the only phenotypic variation is in wing color. These moths all rest on dark-colored tree trunks during the day. Birds prey on the moths by sight in daytime. Our knowledge of the predators' visual system tells us that darker-colored moths are less likely to be seen by the birds. Consequently, those moths less likely to be seen are less likely to be eaten. Finally, the moths that are less likely to be eaten are more likely to leave offspring. Thus we instantiate (RA″) as follows:

Moth $a$ is better adapted than $b$ in (our specified) $E$ iff $a$'s wings are darker colored than $b$'s (in $E$).

(I am here primarily interested in illustrating certain logical points, and hope my view toward the sort of ecological/engineering analysis necessary for complex organisms in complex environments is not overly naive. Lewontin [1978] discusses some of the problems involved. Suffice it to say that although successful ecological/engineering analyses are difficult, they are not impossible.)

With a schematic definition of relative adaptedness, the PNS becomes a schematic law, and with an instantiation of (RA″) we get an instantiation of the PNS. For our moths we get the following:

---

that is, in that $a$, $b$, and $E$ need to instantiated. It is also schematic in the sense that $f(E, \sigma^2)$ is a dummy function and will take different forms in different selection scenarios.

If *a* is darker winged than *b* (in *E*) then (probably) *a* will have
more offspring than *b* (in *E*).

Such an instantiation of the PNS is clearly testable (it has, in fact,
been tested; see Kettlewell 1955, 1956). Moreover, it does what we
want it to do. It explains differential reproduction and thereby explains an important part of the process of evolution by natural selection. (As in this case where we explain the evolution of industrial melanism in a certain species of English moths.)

Earlier I argued that a dispositional property is to be identified
with its causal basis. Notice that the sort of ecological/engineering
analysis described above gives the causal basis for differences in
adaptedness. Taken by themselves the other data, the kind that
come from controlled observations of the actualized fitnesses of
the various types in a common environment, will not necessarily
give us the causal basis for differences in adaptedness. We might
know that darker-winged moths are better adapted than lighter-
winged ones in the environment, but we might not know why.
(The differences in adaptedness may, of course, have nothing to
do with wing color.) Again this is analogous to the situation with
other dispositional properties. From driving two cars under controlled conditions we might know that one was faster than the
other, but we may not know why. Does it have more horsepower,
taller gearing, better areodynamics? From flipping two coins a sufficient number of times we might know that one was biased for
heads, the other for tails; but we may not know why. The instantiations of (RA″), and thus the instantiations of the PNS, require
the causal bases of adaptedness. Therefore, although the data we
can get from controlled observations of realized fitnesses are relevant, we also need some data from ecological/engineering analyses
to produce these instantiations.

To summarize: I have suggested that we give up epistemological
applicability and adopt a schematic definition of relative adaptedness, (RA″). This makes the PNS correlatively schematic and therefore not testable. When we instantiate (RA″) we give up generality
for applicability. Likewise, instances of the PNS become testable
and explanatory, but not general. If I am right, this has ramifications for questions concerning the structure of the theory of evo-

lution by natural selection. Discussion of this topic will be postponed until chapter 4.

## 1.6. OTHER APPROACHES

The approach defended in the last section differs in details from, but is essentially a variant of, what has come to be known as the *propensity interpretation of adaptedness* (or, for those who use the more common terminology, the *propensity interpretation of fitness*). This explication of the notion of adaptedness was introduced into the philosophy of biology literature by Brandon (1978) and Mills and Beatty (1979). It has since been defended by a number of authors, including Burian (1983) and Sober (1984), both of whom suggest, as I did above, that this interpretation of adaptedness is implicit in much work in evolutionary biology. It is my impression that this interpretation is now widely but not universally accepted. Other extant interpretations merit a critical examination. In this section I will deal with two of them. One differs from our approach over the time scale relevant to determining differences in adaptedness. The other differs from our approach both in the definition of natural selection and in the environmental relativity of adaptedness. But both approaches might be considered versions of the propensity interpretation in that neither tries to define adaptedness in terms of actual reproductive success, nor do they try to define it in terms of some specific biological property.

### 1.6a. Long-term vs. Short-term Measures of Adaptedness

Our definition of adaptedness explicitly identifies the adaptedness of an organism with its expected genetic contribution to the next generation. But this approach may be shortsighted. Evolution (with a capital "E") takes place over many generations, so shouldn't we look for some longer-term measure of evolutionary success?

Thoday (1953 and 1958) and, more recently, W. S. Cooper (1984) have argued that we should. Basically, Thoday's definition says that $a$ is better adapted than $b$ if and only if $a$ is more likely than $b$ to have offspring surviving $10^8$ (or some other large number) years from now. Cooper's definition is similar: type *a* is better adapted than type *b* if and only if *a*'s expected time to extinction is greater than *b*'s. (Both authors use the term "fitness" rather than "adaptedness," but both are after a notion that explains natural selection.) If relative long-term performance is really what matters in evolution, then perhaps these notions of adaptedness are more fundamental than our short-term conception.

The problem with Thoday's and Cooper's approaches is that they fail to relate relative adaptedness to the process of natural selection. Put another way, they fail to explain how the process of natural selection can be sensitive to differences in long-term probabilities of surviving offspring. In chapter 3 we will consider the possibility of selection occurring at both higher and lower levels of biological organization other than that of individual organisms, but for the moment let us consider the process of natural selection as it occurs among organisms within a population. That process is the differential reproduction of phenotypes that is due to the differential adaptedness of those phenotypes to a common environment. Evolution results from this process if the phenotypic differences are heritable. How could this process be sensitive to long-term probability (i.e., over many generations) of surviving offspring? Either this long-term probability of surviving offspring corresponds to our short-term expected reproductive success, or it does not. To say that the two correspond means that *a*'s long-term probability of offspring is greater than *b*'s if and only if *a*'s short-term expected reproductive success is greater than *b*'s. If they do correspond, then we have good epistemological reasons for sticking to the short-term measure, since we can operationalize it while there is little hope of operationalizing Thoday's or Cooper's concepts. If they do not correspond, then the long-term probability of surviving offspring is irrelevant to the process of natural selection. That process proceeds through generational time, and organisms with a greater short-term expected reproductive success are probabilistically favored over those with lower short-term expected values. Selection has no foresight; it has no means to discriminate among organisms based on their long-term probability of having surviving offspring.

If species selection (or some other higher-level selection process) is an important process in evolution, then long-term probabilities of survival are relevant, since species generation times will be orders of magnitude greater than organismic generation times. But the probabilities here would apply to species, not to individual organisms or genotypes (see chapter 3). The important point is that when we define relative adaptedness we must have a specific level of selection in mind. We are focusing here on the level of individual organisms within a population. For this selection process the short-term generation measure of adaptedness that we have already defined is the relevant measure.

Indeed, for most purposes the short-term can be equated with a full single generation. But selection is not always so simple. For instance, most aphid species reproduce asexually several times during the summer season. Toward the end of the season, eggs are produced by sexual reproduction and survive through the winter, hatching in the spring. If we measure adaptedness over only one asexual generation, we could get very misleading results, in that one genotype may be a very successful asexual reproducer but a poor sexual reproducer. This genotype might well be at a selective disadvantage when compared with a genotype that is less successful asexually but more successful sexually. To see this we would need to observe the aphids over an entire season that includes several generations. Another example is given by R. A. Fisher's (1930) account of the evolution and maintenance of 1:1 sex ratios. His argument is that in sexually reproducing species the minority sex is always at a selective advantage (because of the greater ease of finding mates). Suppose a population consists of 20% males and 80% females and that an organism of Type 1 in this population has ten offspring, two male and eight female, while type 2 has ten offspring, eight male and two female. Suppose further that chance has nothing to do with this, that this level of reproduction and distribution of males and females is in accord with their differing propensities. If we look at only a single generation, Type 1 and Type 2 seem to have equal adaptedness. But Fisher's argument shows that they will not have equal evolutionary success. The offspring of Type 2 will likely have greater success in finding mates than will the offspring of Type 1. Thus Type 2 is likely to have more grandoffspring than Type 1. To understand this evolutionary scenario we need a full two-generation view of adaptedness.[19]

Nonetheless, the *process* of natural selection operates over short-term generational time; it does not discriminate types based on their probability of surviving $10^8$th generations. Thus the short-term conception of adaptedness is the appropriate one.

## 1.6b. Michod's Approach

Early in this chapter I offered a simple description of the process of evolution by natural selection. The first step in this process is the selective discrimination of phenotypes—by differential survival, differential fecundity, or differential mating ability. When that discrimination is nonrandom, when it is due to differential adaptedness, it is, according to our definition, natural selection. But recall that this process by itself does not result in evolution (i.e., in transgenerational change). For evolution to occur, the phenotypic differences must be heritable. In standard cases this takes place when different phenotypes are associated with different genotypes, and so selection results in the differential replication of genes or genotypes. We called this the second step in the process, or the heritability of phenotypic differences, which depends on the population genetics of the organisms in question. That is, it depends on the genetic system, on the breeding system, and perhaps on other aspects of population structure. In the language of quantitative genetics, the first step is *phenotypic selection*, the second step (or better, the second step followed by the third step, the development of the new distribution of genotypes; see discussion in section 1.1) is the *genetic (or evolutionary) response* to selection (Falconer 1981). Because some authors define natural selection as the combination of these two processes, their definition is equivalent to our definition of evolution by natural selection. For instance, Endler (1986, p. 4) offers such a definition of natural selection. The differences between Endler's approach and mine is largely a matter of terminology. He recognizes the distinctions I have made, in particular the one between phenotypic selection

[19] Sober (1984, pp. 51–58) gives a thorough discussion of Fisher's argument and uses it to make this point.

and evolutionary response. In addition, he realizes that in order to relate to the selection process, fitness must be measured in terms of the relevant within-generation changes (e.g., differential survival). Thus in his approach, as in mine, fitness differences are independent of genetics. But I prefer my terminology to Endler's since my use of "natural selection" refers to the selective process itself, not to the combination of that process with the genetic system. But our differences here are not substantive.

In a series of papers Richard Michod and his associates have elaborated an approach that does differ substantially from my own (Michod 1983, 1984, and 1985; Bernstein et al. 1983; Byerly and Michod 1990). They, like Endler, define natural selection as the transgenerational change that results from phenotypic selection. But unlike Endler, their definition of fitness reflects both selection and genetics.

They follow Fisher (1930) in defining the fitness of a type $i$ (this could be an allele, a genotype, or a phenotype) as the per capita rate of increase in $i$. As mentioned earlier, my definition of adaptedness corresponds to what population geneticists call Darwinian fitness. To understand the difference between Darwinian and Fisherian fitness, consider an example of heterozygote superiority. Suppose at a given locus in a population there are two alleles, $A$ and $a$. Their frequencies are $p$ and $q = (1 - p)$, respectively. If the population is at equilibrium, then the (preselection) genotypic frequencies are $AA$: $p^2$, $Aa$: $2pq$, and $aa$: $q^2$. By definition the Fisherian fitnesses of the three genotypes are all equal at equilibrium. But the Darwinian fitnesses differ, reflecting the within-generation differences in performance of the three genotypes. Assuming the two homozygotes are equivalent, the Darwinian fitnesses are $AA$: $1 - S$, $Aa$: 1, $aa$: $1 - S$. Our definition of adaptedness corresponds to Darwinian fitness in that it is meant to explain the within-generation performance of the various types. What happens to these different types across generations is not decided by differences in adaptedness; that is a matter of population genetics. Thus in our example the heterozygote is better adapted than either of the homozygotes, even though it may not be increasing in relative frequency. To make this example more vivid, let $S = 1$, that is, let the homozygotes be lethal. In that case, even though there is strong phenotypic selection against the homozygotes, each new generation starts with a 1:2:1 ratio of genotypes, and so the Fisherian fitnesses of the three are identical.

Differences in adaptedness, in the sense developed here, explain differences in realized Darwinian fitness. If one thinks that it is differences in realized Fisherian fitness that need to be explained, then one might try to develop a notion of adaptedness that would be explanatory in that context. This is what Michod tries to do. His is a coherent and interesting way of conceptualizing an explanatory theory of evolution by natural selection. But it is potentially misleading and is not the most perspicuous way of representing the process of natural selection. I will offer two criticisms of Michod's approach: one deals with basic strategy, and the second with Michod's execution of that strategy.

Bernstein et al. (1983) introduced an equation termed the *Darwinian dynamic*. The equation represents how the per capita rate of increase of a given type, let us say genotype, is a function of various adaptive capacities of that genotype. Let $X_i$ be the frequency of genotype $i$; $b_i$ and $d_i$ the genotype's intrinsic birth rate and death rate, respectively; and $R$ the resource level. Then,

$$(dX_i/dt)(1/X_i) = b_iR - d_i. \tag{1.1}$$

This equation holds true only for very simple systems where, among other things, reproduction is asexual and the only fitness components are birth and death rates (see Bernstein et al. 1983 or Byerly and Michod 1990 for a detailed list of assumptions). More abstractly, one could rewrite the equation as follows:

$$(dX_i/dt)(1/X_i) = f(A_i, E). \tag{1.2}$$

Equation (1.2) represents the Fisherian fitness of genotype $i$ as a function of various intrinsic adaptive capacities of $i$, $A_i$ and of environmental factors $E$ (Byerly and Michod 1990). As it stands, equation (1.2) is not very general; it does not apply to any system where the genetic system (or aspects of population structure) "mask" the adaptedness of a genotype from its intrinsic rate of increase (Michod 1984). Our example of heterozygote superiority illustrates this point. At equilibrium genotype frequencies remain constant, but the heterozygote is better adapted than either of the two homozygotes. It does not increase in frequency because of the nature of the genetic system ("fair" or Mendelian meiosis and random mating). Thus its intrinsic adaptive superiority over the homozygotes is "masked" by the genetic system (Michod 1984, pp. 258–259, discusses this in greater detail). In such cases, equation (1.2) is inadequate. Here the Fisherian-fitness of a genotype is a function of its adaptive capacities, the environment, and the genetic system. Thus

$$(dX_i/dt)(1/X_i) = f(A_i, E, G), \tag{1.3}$$

where $G$ stands for the genetic system.

Notice that differences in Fisherian fitness imply evolution. To explain differences in Fisherian fitness in terms of differences in adaptedness is to explain evolution by natural selection in those terms. I believe this strategy is unnecessarily complicated. My own strategy is to use differences in adaptedness to explain differences in realized Darwinian fitness, that is, to explain differential reproductive success. As our simple example of heterozygote superiority shows, the ways in which differential reproductive success are translated into evolutionary change can be complicated. This is the domain of population genetics. I think it is helpful to separate the explanation of differential reproduction (the domain of evolutionary ecology) from the multifarious explanations of the evolutionary consequences of differential reproduction. In equation (1.3) $G$ stands for everything we know about population genetics, which, in my opinion, makes (1.3) a bit too unwieldy to be useful.

This criticism is purely strategic. Both Michod and I are interested in explaining evolution by natural selection, and both of us think that relative adaptedness plays an important role in such explanations. My view is simply that phenotypic selection and genetic response to selection are clearly separable phenomena, and, given the complexity of each, they are best studied separately, especially if we are concerned with conceptual issues at the foundation of the relevant explanations.

My second criticism goes deeper. I think that Michod's execution of his strategy is potentially misleading. His approach requires us to be able to identify *intrinsic* properties of genotypes. In equation (1.1) $b_i$ and $d_i$ stand for the intrinsic birth rate and death rate of genotype $i$. In the case of the simplest replicators the ratio $b_i/d_i$ is the adaptedness of type $i$. For organisms with more complicated life histories, adaptedness cannot be so easily analyzed. Nonetheless, according to Michod, the adaptedness of a genotype is an intrinsic property of that genotype. My criticism is that, in the relevant sense, genotypes have no intrinsic properties. In order to make this criticism, a brief excursus into metaphysics is necessary.

### INTRINSIC PROPERTIES

What is an intrinsic property? It is something an object has in virtue of its "inner structure" rather than in virtue of its relation to other objects. For instance, the height of a steel bar is an intrinsic property, whereas its apparent color (or the wavelengths of light reflected from its surface) is not intrinsic since it depends on the distribution of wavelengths in the ambient light. The wavelengths of light reflected from an object depend on two factors: the ambient light and the surface molecular structure of the object. Thus the apparent color of an object is not an intrinsic property, but its surface molecular structure, its "intrinsic color," is.

The apparent color of an object changes in different lightings, but its surface molecular structure remains the same. Perhaps this is the key to the distinction between intrinsic and nonintrinsic properties. If so, the distinction is not an either/or one, but rather one of degree. The surface molecular structure of a piece of steel remains constant over a range of conditions or, to use the terminology that will be relevant to our subsequent discussion, over a range of environments. But it is not invariant across all environments: it will change at high temperatures. Likewise for other properties. Consider height, for example. The height of a steel bar is relatively invariant. Across the range of temperatures we normally encounter, it will change detectably, but not dramatically. It will also vary with the amount of downward pressure exerted on it. But compared to the height of, say, a down pillow, the steel bar's height is relatively invariant across this environmental gradient.

Are there any properties that are invariant over all environments? This might be the case with, for example, the atomic number of gold. Indeed, if a hunk of gold changed in its atomic number it would no longer be gold. But that is irrelevant. What is relevant is whether or not the property of having atomic number

79 is invariant for an object, for a bit of matter, over all conditions. And, of course, it is not. Compared to most elements gold is very stable, but atom smashers and black holes can destroy its atomic structure.

If we accept this account of intrinsic properties, does it follow that there are no intrinsic properties? No. Some properties are more intrinsic than others. The shape of a piece of gold is less intrinsic than its atomic structure because it is less invariant across a range of conditions. But for most practical purposes, the atomic structure of gold and the properties that flow from that structure—for example, its malleability and chemical inactivity—are intrinsic properties of a piece of gold. Its shape is less intrinsic, its apparent color even less. The above is not the only possible account of intrinsic properties, but it is a workable account and is the one relevant to Michod's claims about the intrinsic properties of genotypes.

### INTRINSIC PROPERTIES OF GENOTYPES

In the context of natural selection the relevant properties of genotypes are their phenotypic effects. The nucleotide sequence of a strand of DNA is a genotypic property, but it is relevant to natural selection only insofar as it has phenotypic effects.[20] As is well known, the genotype of an organism does not unambiguously specify the phenotype. The genotype specifies a range of possible phenotypes, often called the *norm of reaction*. The actual phenotype depends on the genotype and the environment in which development takes place. In figure 1.3 the phenotype is plotted against an environmental gradient for genotype $G_1$. How can we determine the phenotypic effect of the genotype? How can we determine the separate contributions of genotype and environment to the phenotype? We might be able to do this by means of a biochemical analysis. For instance, we might know that the genotype $G_1$ produces an enzyme and the enzyme's activity is temperature sensitive, where the enzyme's activity is the phenotypic feature of interest. Thus the effect of the genotype is the enzyme, the effect of the environment is on the level of activity of the enzyme.

This method will usually not be available, nor will it readily

---

[20] See the detailed discussion in chapter 3.

32

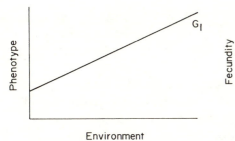

Figure 1.3. Diagram showing how the phenotype produced by a given genotype ($G_1$) can vary across an environmental gradient.

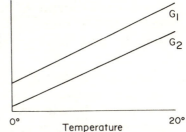

Figure 1.4. Diagram showing how the phenotypes (in this case fecundity) of two genotypes can vary consistently across an environmental gradient (in this case temperature).

yield quantitative answers to our second question. A much more general method, one that will yield quantitative answers, is the analysis of variance. But in order to apply that method to determine the separate contributions of genotype and environment to phenotype we need variation in both genotype and environment. In our case we need a second genotype, $G_2$ (see figure 1.4). Let the phenotype be some component of adaptedness, say, fecundity. Let the environmental gradient be one of temperature. An analysis of variance applied here would allow us to partition the total variance into environmental and genotypic components, and those two would account for the total variance.

Michod's basic idea is that we can express the actual fecundity of an organism as a function of an intrinsic property of its genotype, namely, the intrinsic fecundity, and the environment. Byerly and Michod (1990) suggest that in the simplest cases we can express this functional relationship in the following form:

$$F_i = b_i f(E), \tag{1.4}$$

where $F_i$ is the actual fecundity of genotype $i$, $b_i$ is the intrinsic fecundity of $i$, and $f(E)$ is some function of the environment, in this case temperature. In the case represented in figure 1.4 we can represent this functional relation in a nonarbitrary way if we assign values to $b_1$ and $b_2$, such that their difference corresponds to the genotypic variance (the component of the total variance that is due to genotypic differences). With that constraint and with the constraint that the equation be satisfied, that is, that the data of figure 1.4 fit, our choice of $f(E)$ is nonarbitrary. It represents the environmental variance.

Figure 1.4 represents the easiest case for Michod's program. Now consider figure 1.5. Here the graph for $G_1$ is unchanged, and $G_1$'s fecundity is an increasing linear function of temperature. But $G_2$ does not respond to the temperature gradient in the same way. Its fecundity increases with temperature until 10°C, then it declines with temperature. Still $G_1$ is more fecund than $G_2$ at every temperature, and an analysis of variance would yield a genotypic component of the total variance, which could again constrain the difference between $b_1$ and $b_2$. But while an analysis of variance would yield a genotypic component and an environmental component of the total variance, in this case the two components would not exhaust the total variance. The remainder is called the genotype-environment interaction, or $G \times E$. The $G \times E$ represents the fact that the two genotypes do not respond in the same manner to the temperature gradient, or, to put it another way, the relative fecundity of the two genotypes differs in different environments (i.e., different temperatures). Thus in this case we can-

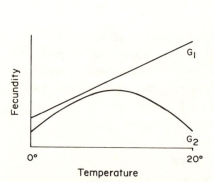

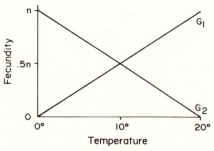

Figure 1.5. Diagram showing a genotype-environment interaction in fecundity, i.e., a change in the relative fecundity of the two genotypes across the environmental gradient.

Figure 1.6. Diagram showing extreme genotype-environment interaction in fecundity. At one end of the temperature gradient $G_1$ is less fecund than $G_2$, while the reverse holds at the other end of the gradient. Here the total variance in fecundity is due to genotype-environment interaction.

not express the fecundity of $G_i$ as the product of $b_i$ and $f(E)$ because the response to the environment of the two genotypes differs. We need different functions, $f_i(E)$, for each genotype $i$. And so we can express the functional relationship of actual fecundity to intrinsic fecundity and the environment as follows:

$$F_i = b_i f_i(E). \tag{1.5}$$

Note that Byerly and Michod (1990) recognize the necessity of indexing $f_i(E)$ to each genotype $i$.

Now consider figure 1.6, which represents the least favorable case for Michod's program. An analysis of variance shows that all of the variance is due to genotype-environment interaction, that is, there is neither genotypic nor environmental variance. Can we still plug in values for $b_i$ and $f_i(E)$ so that equation (1.5) is satisfied? Indeed we can. Let $n$ ($n > 1$) be the maximum fecundity. Then we can let

$F_1 = nf_1(E)$ and
$F_2 = 1f_2(E)$,

where $f_1(E)$ is a linear increasing function of $E$ (i.e., temperature) with initial value $f_1(0°) = 0$ and final value $f_1(20°) = 1$, and $f_2(E)$ is a linear decreasing function of temperature with initial value $f_2(0°) = n$ and final value $f_2(20°) = 0$. This assignment of parameter values makes equation (1.5) fit the data of figure 1.6. So, since $n > 1$, we can conclude that $G_1$'s intrinsic fecundity is greater than $G_2$'s. But we could also let

$F_1 = 1f_1(E)$ and
$F_2 = nf_2(E)$,

where $f_1(E)$ is a linear increasing function of temperature with initial value $f_1(0°) = 0$ and final value $f_1(20°) = n$, and $f_2(E)$ is a linear decreasing function of temperature with initial value $f_2(0°) = 1$ and final value $f_2(20°) = 0$. This fits figure 1.6 equally well. So we can conclude that $G_2$'s intrinsic fecundity is greater than $G_1$'s.

Is $G_1$'s intrinsic fecundity greater than $G_2$'s or vice versa? Clearly in the case of figure 1.6 there is no nonarbitrary answer to this question. Getting equation (1.5) to fit the data is not a problem. The problem is that when there is a large $G \times E$ component in the total variance, the assignment of parameter values is largely arbitrary. In the case of figure 1.6 there is no fact of the matter as to the value of $G_1$'s intrinsic fecundity, and there is no fact of the matter as to the relation between $G_1$'s and $G_2$'s intrinsic fecundity. Put another way, the intrinsic fecundity of a genotype is supposed to be the environmentally invariant contribution of the genotype to actualized fecundity. But when there is a genotype-environment interaction there is no environmentally invariant genotypic contribution to actualized fecundity. For the sake of concreteness I have discussed fecundity, a *component* of overall adaptedness, rather than overall adaptedness. But the above considerations apply with at least as much force to overall adaptedness.

Genotype-environment interactions pose a severe obstacle for Michod's approach. Insofar as they have been empirically investigated such interactions seem to be ubiquitous in nature (see, for example, Sultan 1987; for discussion and further references, see Falconer 1981). I conclude that this approach will not work, that in the relevant sense adaptedness is not an intrinsic property of a genotype.

Perhaps this problem relates to the philosophical obscurity of the notion of intrinsic property. If so, we could possibly avoid the problem by replacing the obscure "intrinsic" with the biologically precise notion of "heritable." This indeed is what Byerly and Michod (1990) have done. But I do not think it helps. (We will discuss the concept of heritability in greater detail in chapter 3.) The crucial fact about heritability is that it is not environmentally invariant. One can define heritability, $h^2$, either phenotypically or genetically. Phenotypically, $h^2$ is the regression of offspring phenotype against parental phenotype (Roughgarden 1979, p. 136). It is a measure of the degree to which offspring resemble their parents with respect to deviation from the population phenotypic mean. For instance, a positive heritability of height would mean that taller than average parents would tend to produce taller than average offspring. Genetically, $h^2$ is defined as the fraction of the total phenotypic variance due to additive genetic effects. The additive genetic variance is that portion of the total genetic variance that is transmitted from parent to offspring.[21] In the case of either

---

[21] See Roughgarden (1979, pp. 152–155). In symbols: $h^2 = V_A/V_T$. $V_A$ is the additive genetic variance and $V_T$ is the total variance in the population. It is the sum of $V_A$, $V_D$, and $V_e$, where $V_D$ is the *dominance variance* (that part of the genetic variance

definition, $h^2$ must be made relative to a particular population and particular environmental conditions.

It is clear that at the least $h^2$ must be relativized to a particular population. Height will be heritable in one population in which there is heritable variation in height. But in another population in which there is no variation in height, or no *heritable* variation in height, height will not be heritable. (Consider a population in which everyone is six feet tall.) Less obvious, but more important for present concerns, is the fact that genetically identical populations can have differing heritabilities for the same trait in different environments. This is so because the environment can control genetic expression. Let us return to the example of figure 1.6. At $0°$ fecundity is heritable, with genotype $G_2$ having higher fecundity than $G_1$. At $20°$ fecundity is heritable, with genotype $G_1$ having the higher fecundity. But what about a population of $G_1$'s and $G_2$'s that exists at $10°$? There fecundity would not be heritable. What is true at $10°$ is also true of a population of $G_1$'s and $G_2$'s spread evenly over a temperature cline of $0°–20°$.[22]

Thus the problem we had in specifying a *nonarbitrary* intrinsic fecundity for either of the two genotypes in the example of figure 1.6 applies equally when we try to specify the heritable fecundities of the two types. In that example fecundity is not heritable in the population as a whole. However, if we restrict our attention to a small range of the temperature gradient, fecundity will be heritable. Heritability, like adaptedness, is not environmentally invariant.

INTRINSIC PROPERTIES VS.
PROPERTIES-IN-ENVIRONMENTS

If adaptedness is not an intrinsic (environmentally invariant) property of genotypes, is it a property of genotypes? This question

---

that is not additive) and $V_e$ is the *environmental variance*, which is simply the residual unexplained variance.

[22] See Clausen and Hiesey (1958), J. P. Cooper (1959), and Clay and Antonovics (1985) for real-life examples of environmental control of genetic expression and thus of the environmental relativity of heritability. Both Clausen and Hiesey (1958) and J. P. Cooper (1959) showed that when certain plants are moved from their native environment to a different environment, new additive genetic variation is exposed. Clay and Antonovics (1985) found greater among-family variation in greenhouse-raised ramets than in those raised in their native field.

divides into two questions. Is adaptedness a property of geno-types *rather than phenotypes*? Is adaptedness a property *rather than a relation*? The former question will be addressed in detail in chap-ter 3, where I will argue that adaptedness is an attribute of phe-notypes, not genotypes. But for the moment let us ignore the ge-notype-phenotype distinction and concentrate on the latter question.

The height of a steel bar varies over temperature gradients and over gradients of differing gravitational forces. Does that mean that height is not a property of the bar, but rather a relation be-tween bar and environment? From a logical point of view one could say so. But logic does not force this decision on us; from a logical point of view it is equally correct to say that having a certain height in an environment is a property of the bar. In the first case, *height-of-2-meters* is a two place relation between object and envi-ronment. In the second case, *height-of-2-meters-in-environment-E* is a property of the object.

If logic does not force the decision, are there any grounds on which to base it? I think there are, but they are purely pragmatic. I have chosen to discuss the height of a steel bar since that seems to be a paradigmatic example of a property. If it is not a property simply because it varies across different environments then, as discussed above, it is doubtful whether there are any properties at all. Under the conditions we normally encounter, the height of a steel bar varies in a barely detectable way. For normal purposes it would be fruitless to insist on thinking of that height as a relation between bar and environment. Thus we think of the height of a steel bar as a property of the bar where that property is implicitly relativized to the environmental conditions we normally encoun-ter.

What about adaptedness? Recall figure 1.6. It illustrates the fact that, in general, the adaptedness of a genotype is not environmen-tally invariant. In figure 1.6 not only does realized fecundity vary with environment, but in this case there is no environmentally in-variant genotypic contribution to fecundity. But suppose we look only at what happens at 20°. There the variance in realized fecun-dity is totally due to genotypic differences. $G_1$ is more fecund than $G_2$ at 20°. If we look at different genotypes in common environments, then the differences in fecundity, or more generally realized fitness, will be attributable to differences in the genotypes. It is to such cases that the PNS applies. (This fact and the concept of environment relevant to such comparisons will be the major topics of chapter 2.) When we compare the adaptedness of genotypes (or organisms) in common environments, we can think of adaptedness as a property of the genotype (or organism). But clearly it is a property that must be relativized to a given environment, that is, it is a property-in-an-environment, and not an intrinsic property. And so the relation of one genotype (or organism) being better adapted than another must also be relativized to a given environment.

Relative adaptedness, unlike the height of a steel bar, does vary significantly across environments. When we fix the environmental context we can think of adaptedness as a property of a genotype (or organism). But it is a property only modulo a certain environment. That is, it is a property-in-an-environment. In the same way, we can think of adaptedness as a relation between genotype (or organism) and environment. But for the purposes of explaining natural selection we must compare the adaptedness of different entities within a common environment. Hence the utility of thinking of adaptedness as a property-in-an-environment. And, hence the disutility of thinking of adaptedness as an intrinsic or environmentally invariant property.

## 1.7. Adaptation: Process and Product [23]

Having defined the notion of relative adaptedness and having explored its role in the theory of natural selection, I now want to look at two cognate notions. The word "adaptation" is used in two quite distinct ways. It is sometimes used to refer to a process and sometimes to certain sorts of traits of organisms. I will argue that both of these notions are explicable in terms of relative adaptedness.

Two processes can result in organisms being adapted to their environment.[24] One is an evolutionary process, the other an on-

---

[23] This section is taken with some modification from Brandon (1981a, section 3).
[24] There are two, but not only two. Another nonevolutionary process by which

togenetic process. The ontogenetic process can take many forms. It can be an obligate (hard-wired) process whereby an organism changes in response to environmental changes. For instance, frogs of the species *Hyla versicolor* change their color in response to the color of the background on which they are sitting. At another extreme, humans and other animals employ various learning rules to adapt their behavior to environmental conditions. This process, often called *adaptability*, should not be confused with the evolutionary process of *adaptation*. Adaptation can be identified with the process of evolution by natural selection. It is by this process that populations change in response to environmental changes, and it is this process that produces what Darwin called "organs of extreme perfection and complication" (1859, p. 186).

What is an adaptation? There are basically two extant answers. One, a minority view, holds that any trait that increases the adaptedness of its possessor is an adaptation (see Munson 1971; Ruse 1971; Nagel 1977; and Bock 1980). I will call this the *ahistorical concept of adaptation*. Although the sorts of physiological, biomechanical, behavioral, and ecological studies that could justify the claim that a trait is an adaptation in this ahistorical sense are immensely interesting, the concept itself is of no importance in evolutionary theory. We already have the ahistorical notion of relative adaptedness; it plays a central role in the theory of natural selection. But simply to say that some trait benefits its possessors is to say nothing about its evolutionary history or future, and so it says nothing of relevance to the theory of evolution by natural selection. This irrelevance presumably accounts for the view's minority status.

The majority view is that an adaptation is the product of the process of evolution by natural selection (see G. C. Williams 1966; Lewontin 1978; Brandon 1981a; Gould and Vrba 1982; and Sober 1984). I believe that this view is essentially correct but needs some clarification. Whether adaptations are properly considered to be features of genes, genotypes, or phenotypes is a question we will deal with in chapters 3 and 5. For now I will assume the position I will defend there, that is, that adaptations are features of phenotypes. We could talk of whole phenotypes as adaptations, but

organisms can "fit" their environment is the competitive sorting of differing phenotypes across differing environments (Horn 1979).

such talk would be uninformative since, presumably, all pheno-types are to some extent the products of the process of evolution by natural selection. If instead we talk about parts of whole phe-notypes, that is, phenotypic traits, as adaptations, then such talk can be informative because not all traits are adaptations. To say that a trait is an adaptation is to say something about its causal history. Not all traits have the requisite sort of causal history to be adaptations.

It should be obvious that there are many examples of traits that are not gene linked at all and so are not adaptations. These need not concern us. There are, however, two interesting categories of traits that are not adaptations. They are (1) epiphenomenal traits, and (2) traits due to chance. Let us consider the latter first. A trait may appear with one mutation. Such a trait is not an adaptation regardless of its effect on the adaptedness of its possessor. If the trait does increase the adaptedness of its possessors and thereby increases in frequency in the population it will then become an adaptation. But its initial appearance is explained by the chance mutation. Thus at that point it is not an adaptation. More interest-ingly, traits may actually evolve, that is, be the product of an evo-lutionary process, and still not be adaptations. The process is evo-lution by random drift. As discussed above, selectively neutral traits can become fixed in a population by that process. Further-more, as our basset-shepherd case illustrates, even traits that are disadvantageous relative to their alternatives can become fixed by this chance process in a small population.

Traits due to chance are not just logical possibilities. In particu-lar, if the generally accepted theory of speciation is correct, then random drift in small populations has had tremendous evolution-ary effect.[25] Indeed it is likely that some differences among species are largely due to chance and that natural selection just fine-tunes these given differences to fit local environments.

I call the second category of traits that are not adaptations "epi-phenomenal traits." This is a rather heterogeneous category. It in-

[25] By what Mayr calls the *founder principle*. New species are formed from small geographically isolated founder populations. The gene pool of this population is, to a large extent, a random sample of the original gene pool. Also, the small size of the founder population increases the subsequent role of random drift. See Mayr (1963).

cludes traits that have evolved not on their own merit, but due to their connection to other evolving traits. The two simplest types of connections are *gene linkage* and *pleiotropic connections*. When a not particularly deleterious gene is closely linked on a chromosome to a gene being strongly selected for, it can hitchhike its way to prominence. Pleiotropic connections are similar but more permanent. Pleiotropic genes are genes that affect more than one part of the phenotype (or more than one trait). For example, a gene may code for an enzyme that helps detoxify a poisonous substance common in its environment.[26] It does so by converting the poison to an insoluble pigment. Two new traits become common in the population over generational time. One, the ability to handle the toxic substance, is an adaptation. The other, the resultant color of the organisms having the first property, is not an adaptation. Its presence is not explained by what it does but rather by its pleiotropic connection to an adaptation.

What is the basis of this distinction? In the given environment there is a causal relationship between the ability to handle the toxic substance and adaptedness; there is no such causal relationship between color and adaptedness in *this* environment. (Of course in another environment color might have a close causal connection with adaptedness.) To assert that something is an adaptation is to make a causal-historical statement.

A more nebulous type of epiphenomenal connection is illustrated by the following examples. A beating heart makes sounds. Although nowadays heart sounds are a diagnostic aid and thus beneficial to the people producing them, it is highly doubtful that heart sounds played any role in the evolution of the human heart. Thus we do not want to call the making of heart sounds an adaptation. Humans and other organisms would probably have been just as well off with an organ that circulated blood silently. (Such an organ is conceivable, but probably not an open possibility for any of the organisms in question. But if that is so, then we can hardly say that an organism would be just as well off without heart sounds: it would not be as well off, but would be dead.) Consider a second example. Humans, some of them at least, have the ability

[26] This example is from Lewontin (1978, p. 228), where he discusses more fully some of the topics mentioned here.

to do meta-logic. Critics of evolutionary approaches to epistemology might claim that this ability has had no possible evolutionary significance and so no evolutionary explanation of its presence can be given. But it may be that the ability to do meta-logic is an epiphenomenal by-product of a generalized intellectual apparatus that is a human adaptation. (The epiphenomenal connections illustrated in the two preceding examples are less objective and more a product of how we decide to describe phenotypic features than gene linkage or pleiotropic connections. But this is irrelevant to present concerns.)

We have seen that not all traits are adaptations, that is, not all traits have the same type of causal history. It follows from our discussion that, in contrast to the ahistorical notion of adaptation, the condition that a trait be beneficial to its possessor is neither a necessary nor sufficient condition for it to be an adaptation. It is not a necessary condition because, due to environmental changes, a trait that once increased the relative adaptedness of its possessors (and increased in relative frequency because of that) may now be neutral or even deleterious. Likewise it is not a sufficient condition because traits that once were neutral or even deleterious may now serve to increase the relative adaptedness of their possessors. For example, in our hypothetical case of a pleiotropic connection between poison detoxification and darker color, the ability to detoxify the poison was the adaptation while the darker color had no effect on relative adaptedness. But the ecological situation might change so that the darker color increases relative adaptedness (say by camouflaging the organisms from a newly introduced predator). At that point the dark color is not an adaptation. Gould and Vrba (1982) call such traits *exaptations*. If there is selection for an exaptation it may in time become an adaptation.

The bulk of this chapter has been devoted to developing an adequate definition of relative adaptedness. Having done that it is easy to use that definition to explicate the related concepts of the process of adaptation, that is, the process of evolution by natural selection, and of the products of that process, adaptations. The latter concept is, from a purely conceptual point of view, unproblematic. But there are serious epistemological problems concerning its application. For instance, in the case of pleiotropy discussed earlier there are obvious experimental procedures we could use to determine if the ability to detoxify the poison or the resultant darker color was causally responsible for the increased relative adaptedness of the possessors of that gene. But once we have concluded that it is the ability to detoxify the poison that is causally relevant to increased relative adaptedness *in the present environment,* on what basis do we conclude that the ability to detoxify the poison is the adaptation? Clearly, we need to extrapolate from the present selective environment to past environments. But what evidence can we have for such extrapolations? How reliable are they? These practical epistemological problems and their ramifications will be considered in chapter 5.

# The Concept of Environment in the Theory of Natural Selection

During the last few decades our understanding of natural selection has been advanced by the elimination of certain phenomena from the category of natural selection. Although there are many who have defined natural selection as mere differential reproduction (see Beatty 1984, p. 191, for examples), it is now abundantly clear that such a definition is inadequate. If random drift is an alternative to natural selection and if it can result in differential reproduction, then not all cases of differential reproduction are cases of natural selection. As in the example discussed in chapter 1, lightning may strike vigorous individuals, leaving less well adapted individuals to survive and reproduce. This point has been widely recognized by philosophers of biology, especially those adopting the propensity interpretation of adaptedness. According to that interpretation, adaptedness is a probabilistic disposition, so that greater relative adaptedness regularly leads to greater actualized fitness, but chance can intervene and break that regular connection (see discussion in chapter 1).

The differential reproduction resulting from random drift is not natural selection. Are there other cases where differential reproduction is not natural selection? One such case will receive extensive discussion in chapter 3, but briefly, it involves the differential survival and reproduction of entities at various levels of biological organization. For instance, groups of organisms may survive and send out propagule groups at different rates. Is this group selection, that is natural selection at the level of groups? Since G. C. Williams's (1966) critical examination of this issue, biologists and philosophers of biology have reached near universal agreement that such differential reproduction *need not be* natural selection at the level of groups. In the useful terminology of Vrba and Gould (1986), not all sorting (i.e., differential reproduction) is selection.

For instance, two sister species may differ in that members of one but not the other are resistant to a newly introduced parasite. Because of their lack of resistance, members of the second species have a greatly lowered reproductive rate and eventually the species goes extinct. Members of the first species are unaffected, and in time that species may speciate. Here processes at the level of individual organisms account for the differences in the fates of the two species. This is sorting at the species level, but not species selection.

In chapter 1 I argued that the Principle of Natural Selection (PNS) applies to those cases where the differences in reproductive success are caused by differences in adaptedness to a *common* environment. In such cases differences in some trait or traits lead to differences in the organisms' ability to survive and reproduce; and that in turn leads, in a probabilistic fashion, to differences in actual reproductive success. If the traits are heritable, this leads to a change in the distribution of traits in the next generation, with the better adapted traits being increased in relative frequency. This is the process of adaptation, that is, the process by which organisms adapt to their local environment. If all this is correct, then there is another way, until now unrecognized, in which chance can intervene to disassociate adaptedness from actualized fitness. By chance different types can be unequally distributed over a heterogeneous environment. For example, consider two seed types, *a* and *b*, distributed by wind over a patchy field, some patches containing fertile soil and others toxic soil. By chance a disproportionate number of *a*'s land in fertile soil. Because of that the relative frequency of *a*'s in the field as a whole increases.[1] But this is not natural selection. Type *a*'s have a greater realized fitness than *b*'s, but it is the properties of the environment rather than the properties inherent in the plants themselves that explain their differential success. Recall from our discussion in chapter 1 that adaptedness is a property-in-an-environment, not an environment-independent or intrinsic property. It follows that in order to explain differences in realized fitness in terms of differences in adaptedness one

---

[1] This case will be discussed in greater detail below. One might want to compare my analysis of this case with Beatty's (1984, pp. 192–196) discussion of a similar case. I believe Beatty's analysis is flawed precisely because it lacks the concepts developed in this chapter.

must compare organisms in common environments. Because of this, the concepts of adaptedness (expected fitness) and environment are inextricably linked; one cannot make sense of one independently of the other. Thus if the theory of natural selection is to be an explanatory and predictive theory, rather than a post hoc description of actualized fitness differences, we must be able to determine when organisms are inhabiting common environments and when they are in different environments. This issue is the focus of this chapter.

## 2.1. THREE CONCEPTS OF ENVIRONMENT

What is the environment of an organism (or of a population or species)? The common-sense answer is that it is the sum total of the factors, both biotic and physical, *external* to the organism that influence its survival and reproduction. Indeed, this is the operative conception in ecology. The environment is measurable by means of instruments and analyses that quantify some factor or set of factors that exist independently of the organism in question. For instance, in a long narrow field we might measure the concentration of molybdenum in the soil (figure 2.1). Such measurements can be performed without any involvement of the organisms of interest. The resulting measurements of molybdenum levels give us the scale of environmental heterogeneity with respect to molybdenum. Of course, levels of molybdenum concentration may have no effect on the organisms of interest. If not, we can go back and measure something else. Presumably, the more we know about the biology of the organisms the better able we will be to pick external factors that are indeed relevant. Such methods measure what I will call the *external environment*.

No good ecologist would consider randomly picking some external factor, say molybdenum, and measuring it. We are interested only in those factors of the external environment that affect the organisms (or populations or species). An alternative and more direct method is to use the organisms themselves as measuring instruments.[2] Suppose that we are interested in a particular species

---

[2] The use of organisms as measuring instruments or as "phytometers" was pioneered by Clements and Goldsmith (1924). Recently it has been used extensively in ecological genetic studies by Janis Antonovics and his associates. See Antono-

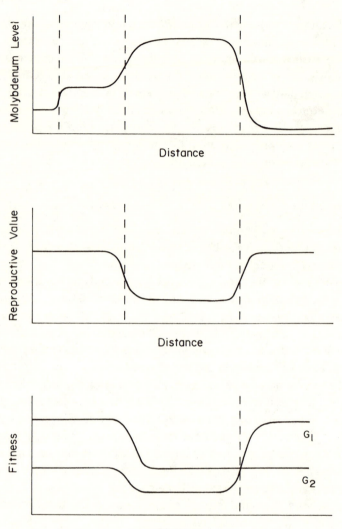

Figures 2.1, 2.2, and 2.3. Schematic illustration of how external (top), ecological (middle), and selective (bottom) environments may show different scales of heterogeneity.

of grass in our field. Suppose further that it is technologically feasible for us to clone seeds in this species. We could clone multiple copies of some particular genotype and plant the resultant cloned seeds out along the distance of the field. Over a period of years we could measure each seed's actualized fitness, or its reproductive value *sensu* Fisher (1930). The result would be something like figure 2.2. Here we are measuring the scale of environmental heterogeneity in terms of the demographic performance of individuals. It follows that, in contrast to the external environment, the scale of heterogeneity that is present depends on the organisms used as measuring instruments (Antonovics, Clay, and Schmitt 1987). (Note that we would want to use multiple copies of the same genotype in measuring this environmental heterogeneity in order not to confound environmental differences with genotypic differences. Alternatively, we could sample from among the genotypes extant in the field, clone copies of each, and plant multiple copies of each genotype along the distance of the field. Then we could average over all genotypes.) Such methods measure what I will call the *ecological environment*. The ecological environment reflects those features of the external environment that affect the organisms' contributions to population growth.

In the present context we are primarily concerned with the concept of environment as it functions in the theory of natural selection. Selection requires differential reproduction, that is, different "types" must make different contributions to the next generation. (I will discuss shortly whether these "types" are genotypes or phenotypes, but for the moment let's consider genotypes.) What I will call the *selective environment* is measured in terms of the relative actualized fitnesses of different genotypes across time or space. For instance, in our field, instead of planting multiple copies of a single genotype, we would plant multiple copies of two or more genotypes. The result would be figure 2.3. The scale of heterogeneity reflects the differential performance of genotypes in different regions or at different times, that is, genotype-environment interactions.[3]

---

vics, Clay, and Schmitt (1987) and Antonovics, Ellstrand, and Brandon (1988). Also see Turkington and Harper (1979) and Turkington et al. (1979).

[3] The distinction of external, ecological, and selective environments is introduced in Antonovics, Ellstrand, and Brandon (1988).

## 2.2. DEVELOPMENTAL VS. SELECTIVE ENVIRONMENT

The primary focus of this chapter is on the concept of selective environments. I will argue that this concept is central to the theory of natural selection. However, one might think that within evolutionary biology there are at least two distinct notions of environment, one concerned with development, the other with selection. In chapter 1 I described the process of evolution by natural selection as a three-step process. One step is the selective discrimination of different phenotypes. This process occurs within an environment, that is, the relative adaptedness of the phenotypes depends on the particular environment. The next step is the differential replication of the underlying genotypes. In this way the genotypic distribution of the next generation is changed. The final step is the development of this new set of genotypes. This too occurs within an environment, that is, the phenotype produced by a given genotype depends on the environment in which it develops. Thus the entire three-step process depends on developmental and selective environments, and these two types of environments can be distinguished in terms of the different roles they play. If this distinction could be maintained it would simplify the present investigation. If phenotypes were static entities, that is, if they could be identified with the end result of development, then the distinction would hold.

But from the point of view of natural selection, phenotypes are dynamic, temporal entities. For instance, in some circumstances the selectively relevant differences may be differences in the time at which adult morphology develops, not differences in the final morphology. (An example of this would be tadpoles that develop in temporary ponds. When the pond dries up, those who have developed adult morphology can hop away, those slower to develop will die.) More generally, treating phenotypes as static products of development rather than treating developmental processes as parts of the phenotype commits one to clearly unacceptable views. For example, consider the following: In one environment, $E_1$, genotypes $G_1$ and $G_2$ produce the same distributions of heights. In $E_2$ they produce different distributions, with $G_2$'s distribution being shifted to smaller heights (figure 2.4). Within both environ-

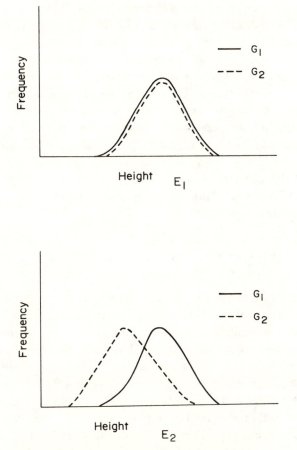

Figure 2.4. Diagram showing how the relative phenotypic expressions of two genotypes can differ in different environments. This illustrates how environmental effects on development are a part of the selective environment.

ments there is directional selection of equal intensity for increased height, that is, the regression of fitness on height is the same in both environments. To treat phenotypes as fixed products, to ignore developmental processes in thinking about selective environments, commits one to the view that environments $E_1$ and $E_2$ are identical selective environments. However, it should be clear that they are not. Suppose the difference is due to the following: $E_1$ contains a substance lacking in $E_2$; that substance promotes growth; and $G_1$ can synthesize this substance from elements available in $E_2$ while $G_2$ cannot. Clearly, the presence or absence of the substance makes a selective difference: in $E_1$ the two genotypes are selectively equivalent, in $E_2$ they are not. That is not to say that genotypes are directly exposed to selection. In $E_2$ $G_1$'s ability to synthesize the substance is selectively favored. But that, of course, is a feature of $G_1$'s phenotype (in $E_2$). $E_1$ and $E_2$ are different selective environments. They are different (relative to $G_1$ and $G_2$) because of the differential effects they have on the development of the two genotypes. Thus the developmental environment is part of the selective environment.

## 2.3. ENVIRONMENTAL HOMOGENEITY AND HETEROGENEITY

Thus far I have argued that it is important for us to be able to distinguish populations inhabiting heterogeneous environments from those inhabiting a single homogeneous environment. But what, exactly, does it mean to say that a selective environment is either homogeneous or heterogeneous? When an environment is homogeneous with respect to selection, within that environment different copies of the same "type" will do equally well with respect to selection, that is, they will have the same relative expected reproductive success. Since phenotypes directly interact with the environment, the "types" that are relevant are phenotypes. (See chapter 3.) But for the reasons discussed above, we must not take an overly simplistic view of phenotypes; we must remember that phenotypes are temporally dynamic entities. As we will see below, the ideal test of homogeneity would consist of distributing multiple copies of the same genotype over an area, allow them to grow, and observe their realized fitnesses. Thus when I want to stress the fact that selection involves the interaction of phenotype and environment, I will describe homogeneity in terms of the performance of phenotypes. When I am concerned with the experimental measurement of heterogeneity and homogeneity, I will talk of the performance of genotypes. (But I will assume that there is a class of phenotypes produced by the genotypes and that selection acts on phenotypes, i.e., that there is heritable variation in fitness.) When neither needs to be stressed, I will simly refer ambiguously to "types." None of this should cause confusion so long as we remember that it is the dynamic phenotype that is directly exposed to selection.

To begin I will explicate the notions of homogeneity and heterogeneity as they apply to ecological environments; then I will extend this analysis to selective environments. An area is an ecologically homogeneous environment if within that area different copies of the same type have the same expected reproductive value, that is, the same level of adaptedness. This is strictly analogous to von Mises's definition of *randomness*, according to which a sequence is random with respect to a certain outcome if and only if there is no subdivision of the sequence independent of the outcome, such that the probability of the outcome is different in the subsequence from that in the whole sequence (see Salmon 1971). For instance, consider a sequence of tosses of a coin. That sequence is random with respect to the outcome (heads or tails) if and only if there is no subdivision of the sequence independent of the outcome in which the probability of heads differs from that in the whole sequence. In other words, the permitted subdivisions are those that can be effected independent of the possible outcomes, for example, odd numbered tosses, tosses on Thursdays, tosses in ambient temperature 10°C or below. But, to divide the sequence into those tosses that yield heads and those that do not is not an admissible subdivision. So the sequence is random if and only if there is no admissible subdivision that affects the probability of the outcome. Note that this is not an epistemic concept; it is not defined in terms of known or knowable admissible subdivisions. It is easy enough to define "epistemic randomness." A sequence is epistemically random if and only if there is no *known* admissible subdivision that affects the probability of the outcome.

In the case of ecological environmental homogeneity, an area or population is homogeneous if and only if there is no admissible subdivision of the area or population in which the adaptedness of a given type differs from its adaptedness in the whole area or population. Of course we do not expect every copy of the same type to have the same realized fitness. As mentioned above, selection is not a deterministic process. But because the outcome in question is realized fitness, we cannot subdivide the population on that basis. Figure 2.5 illustrates two admissible and relevant subdivisions. In figure 2.5a realized fitness is plotted along a spatial transect. Clearly if we divide the population in two at the 10-meter mark, the two subpopulations are more homogeneous than the original population. Indeed this can be quantified. The variance in the realized fitnesses for a given type can be used as an inverse measure of homogeneity; the lower the variance, the greater the degree of homogeneity. Figure 2.5a is what we might expect for sessile organisms with discontinuous environmental variation. But sometimes environmental variation may not be along a spatial (or temporal) scale. Figure 2.5b illustrates a case where organisms form groups, and fitness of type *a* depends on its relative frequency within the group. Here realized fitness is plotted against the relative frequency of *a* within the group. Again we can partition the population such that the variance in the fitness of *a* within each subdivision is less than the variance in the fitness of *a* within the whole population. The difference between the two cases is significant. It is probably natural to think of the partitioning of the biological world into homogeneous environments as being a spatial, or spatiotemporal, division. But figure 2.5b shows that this need not be the case. (We may imagine that the groups of organisms depicted in figure 2.5b are motile and have highly overlapping ranges.)

Having said that homogeneity is inversely related to the variance of the realized fitness of a given type, I should point out that we are not likely to get the data we need to correctly partition heterogeneous populations into homogeneous subpopulations from simple observations of natural populations. In all likelihood some sort of perturbation study will be needed. For instance, suppose figure 2.5a represents a species of grass in a field. As discussed earlier, in this case the ideal experiment would be to clone multiple copies of seeds with the relevant genotype and then plant them along the 20-meter transect. (If it were not technologically feasible to clone seeds, then we could use cloned individuals and get less complete data—data that ignore processes occurring at germination and at the seedling stage. Or we could use sibships of seeds,

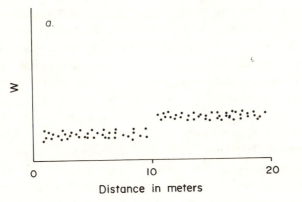

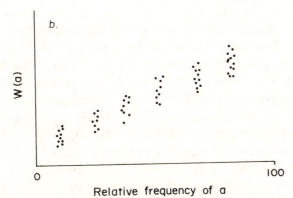

Figure 2.5. Hypothetical data points of realized fitness plotted against environment. In (a) realized fitness points are plotted along a spatial transect. Here we can divide the transect at the 10-meter mark, resulting in two sets of data points both more homogeneous with respect to fitness than the original set. In (b) realized fitness is plotted against the relative frequency of type *a*. Again this data set can be subdivided in a way that increases the homogeneity with respect to fitness.

that is, seeds gathered from one plant.) Over a period of time we would measure the realized fitnesses of these plants (by keeping track of mortality, number of inflorescences, or whatever else might be relevant). That would give us the data points of figure 2.5a. This method has been called the *phytometer* method (Clements and Goldsmith 1924). As the name implies, in this method plants are used as measuring instruments of environments. There are many advantages to this method (see Antonovics, Ellstrand, and Brandon 1988); but the primary theoretical advantage is that we are trying to measure how the organisms perceive the environment, and this method provides the most direct possible measure of this. After detecting patterns of environmental heterogeneity we might, as ecologists, want to discover the external factors that correlate with this pattern, for example, concentrations of nitrogen, surrounding vegetation, or whatever. But that information is not needed to ascertain patterns of environmental heterogeneity. Although this method is easiest to apply using sessile plants, it can certainly be used with animals as well. For instance, the data for figure 2.5b might be gathered by distributing groups of salamanders with varying frequencies of *a* into separate ponds.

Thus far we have discussed environmental homogeneity with respect to a single type (ecological environmental homogeneity). But our interest is in selective environments, and selection cannot occur without variation. So we must extend this analysis to populations with multiple types. Figure 2.6 represents the simplest case. Figure 2.6a corresponds to figure 2.5a with the data points replaced by a line fitted to them. Let this be the fitness of genotype $G_1$. In figure 2.6b the fitness of a second genotype, $G_2$, is added. Its fitness also shifts at the 10-meter mark. Thus both 10-meter sections are homogeneous with respect to $G_1$ and $G_2$, that is, within both sections the expected fitnesses of both genotypes are invariant. Of course, each 10-meter section can be divided further, but note that within any such partition the expected fitnesses of the two types remain unchanged. Thus such partitions are irrelevant. To rule them out we can stipulate that a selective environment with respect to types $a_1, a_2, \ldots, a_n$ is the largest partition that is homogeneous with respect to types $a_1, a_2, \ldots, a_n$. Thus the transect in figure 2.6b consists of two selective environments with re-

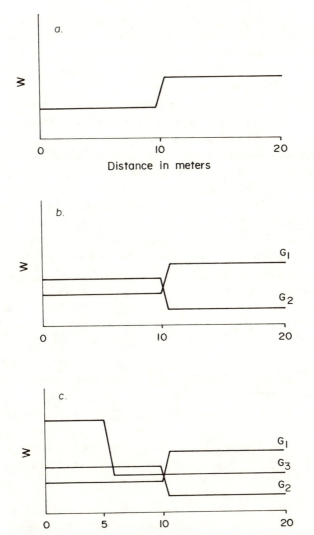

Figure 2.6. Schematic diagram illustrating the concept of selective environmental homogeneity. In (a) the fitness of a given genotype is plotted against distance. In (b) a second genotype is added. Here both genotypes change in fitness at the 10-meter mark, resulting in a change in their relative fitness. Both the first 10-meter section and the second one are selectively homogeneous. In (c) a third genotype is added, resulting in three selectively homogeneous sections.

57

spect to $G_1$ and $G_2$. In figure 2.6c a third genotype is added. $G_3$'s fitness shifts at the 5-meter mark; so with respect to genotypes $G_1$, $G_2$, and $G_3$ we have three selective environments. The first is contained in the first 5 meters, the second in the second 5 meters, and the third in the final 10 meters.

As should be clear from the above discussion, homogeneity is always relative to a set of types. When dealing with natural populations, the relevant types are those that actually occur in the area of interest. It follows that we can make no absolute statements about patterns of selective environmental heterogeneity; the introduction of new types (e.g., by mutation or migration) can change that pattern. For this reason selective environments have no existence independent of the organisms exposed to selection.

In this section I have discussed cases where the environment varies in a discrete manner. This is the simplest case. Moreover, there are many ecological situations where it is realistic to suppose that environmental variation is discrete.[4] But it is also likely that there are other cases where the environmental variation is more or less continuous, as shown in figure 2.7. Here the fitnesses of three genotypes are plotted against a spatial transect. How can this transect be divided into selectively homogeneous environments? One option might be to insist that each point on the transect is a unique selective environment, but this is not a very practical option. Furthermore, it is highly doubtful that any experimental design of the sort we have been discussing would show statistically significant differences between points separated, say, by one centimeter.[5] A better option is to adopt a concept of *selective environmental neighborhoods*, whereby neighborhoods are centered around arbitrarily

[4] For instance, if selection is frequency dependent and the frequency-dependent effects are limited to small discrete groups, then one would expect the environmental variation to be discrete. Another example would be one where the selective environment of a given species depends on the distribution of both predator and prey species and where that distribution differs between small waterfalls in a mountain stream.

[5] Since these differences would be small we would need a large number of replicates to detect them reliably. But there is an obvious physical limitation on the number of seeds that can be planted along a one-centimeter line; furthermore, such dense planting may well introduce density-dependent effects not present in the undisturbed field. These reasons are pragmatic and epistemological. There are also good ontological reasons for rejecting this option, which I will discuss in section 2.5.

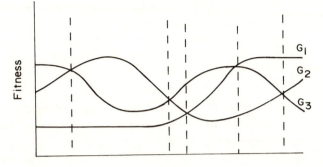

Figure 2.7. Schematic diagram illustrating the concept of selective environmental neighborhoods. This concept is applicable when fitness varies across environment in a continuous rather than discrete way.

chosen individuals. The size of the neighborhood would depend on how much divergence we allow from the relative expected fitnesses of the relevant genotypes at the point of that individual.

There is one constraint that needs to be placed on this method of dividing a continuously varying environment into selective neighborhoods. We should stipulate that within any selective neighborhood the ordinal relation of the relevant genotypes does not change. Graphically this amounts to the requirement that fitness lines not cross within any selective neighborhood. In figure 2.7 the regions between crosses are separated by vertical dashed lines, thus there are a minimum of six selective neighborhoods. The motivation for this constraint is obvious. It hardly makes sense to call an area homogeneous with respect to selection if at one end of the area genotype $G_1$ is favored over $G_2$ while at the other end $G_2$ is favored over $G_1$. Such selective heterogeneity is indicated empirically by a negative genetic correlation in fitness across the ends of the area or by a genotype-environment (in this case spatial position) interaction effect that is greater than the genotype main effect (see Via and Lande 1985; and Antonovics, Ellstrand, and Brandon 1988 for further discussion). With this restriction in place I believe that the concept of selective environmental neighborhoods adequately handles cases where fitnesses vary continuously over the environmental scale. But for ease of exposition I will, for the most part, discuss the simpler cases where there are discrete selective environments.

## 2.4. HABITAT CHOICE VS. CHANCE DISTRIBUTION

Earlier I argued that one way that chance can intervene to dis-associate adaptedness from actualized fitness is by the unequal distribution of types into a heterogeneous environment. Consider the following case: Two genotypes, $G_1$ and $G_2$, are distributed into two discrete environments, $E_1$ and $E_2$. $G_1$ is fitter than $G_2$ in both environments, and both genotypes are fitter in $E_2$ (see figure 2.8). But the two genotypes are not distributed equally into the two en-vironments; $G_2$ is much more abundant in $E_2$. Let $N$ be the number of copies of each genotype in generation 1 and let the fitnesses be $W(G_1,E_1) = 1$, $W(G_1,E_2) = 2$, $W(G_2,E_1) = 0.8$ and $W(G_2,E_2) = 1.8$ (where $W(G_i,E_j)$ is the fitness of the $i^{th}$ genotype in environment $j$). Let $0.8N$ of the $G_1$'s live in $E_1$ and $0.8N$ of the $G_2$'s live in $E_2$. Then the overall fitnesses are:

$$W_0\,(G_1) = (0.8N{\cdot}1 + 0.2N{\cdot}2)/N = 1.2 \tag{2.1}$$
$$W_0(G_2) = (0.2N{\cdot}0.8 + 0.8N{\cdot}1.8)/N = 1.6, \tag{2.2}$$

[where $W_0(G_i)$ denotes the overall fitness of genotype $i$]. In this case, although $G_1$ is better adapted than $G_2$ in both environments,

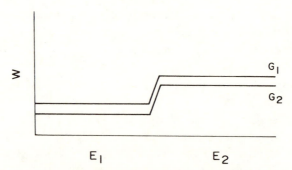

Figure 2.8. Diagram showing the relative fitnesses of two genotypes in two environments. Here $G_1$ is fitter than $G_2$ in both environments, and both genotypes are fitter in $E_2$ than they are in $E_1$.

$G_2$'s overall (realized) fitness is higher. Those familiar with group selection models for the evolution of altruism will recognize the formal analogy the above bears to such models. An altruistic type can increase in relative frequency even though within each group it is less fit than its nonaltruistic competitors. It can do so only if there is variation in the relative frequency of altruists between groups and if the groups with higher percentages of altruists do better than groups with lower percentages (this will be discussed in chapter 3). (Indeed the above case might fall under some definitions of group selection, e.g., that presented by Wade 1984.) $G_2$'s overall fitness is higher only because the richer environment happens to contain more $G_2$'s.

What are the evolutionary consequences of this? $G_2$ will increase in relative frequency by the end of generation 1. If, as by hypothesis, the two types are randomly distributed over the two environments, then the next generation may find $G_1$ the lucky type (i.e., the type disproportionately represented in $E_2$). Like random drift, this process may have long-term evolutionary consequences; but the consequences will not be directional like those of evolution by natural selection (see Gillespie 1978).

Note that in the above example the overall fitness values are potentially misleading. From them we might conclude that there was selection for $G_2$. But there was selection for $G_2$ only in the tautologous sense that $G_2$ increased in relative frequency. In that sense the "theory of natural selection" is no theory at all; it has no predictive or explanatory value. In the above case the overall fitness values are not extrapolatable because the genotypes are randomly distributed over the heterogeneous environment each generation. Thus the overall fitnesses are not heritable and so cannot be used to predict evolutionary changes in future generations. But we cannot make the converse inference. We cannot conclude from the extrapolatability of fitnesses that selection is occurring in a homogeneous environment. The unequal distribution in the above case could have resulted from a rare event (e.g., a 100-year flood). If so, the two types may stay unequally distributed for a number of generations, and so the overall fitnesses may be extrapolatable over some number of generations. Still, this is selection in heterogeneous environments. In both the original example and in this modified case the causal process of selection takes places in $E_1$ and $E_2$, and in both environments $G_1$ is selectively favored (i.e., better adapted). The final part of the causal story concerns the distribution of the two types over the two environments.

How do we know that the distribution is the result of chance? Suppose our two genotypes are from a species of phytophagous insects and that the two environments correspond to two different plant species co-occurring in a field. Larvae feed and develop on the plant on which they were hatched, and selection occurs during this stage. In this case it is the behavior of the egg-laying females that distinguishes random distribution from habitat choice.[6] We could experimentally determine whether or not $G_1$ females preferred $E_1$ plants and $G_2$ females preferred $E_2$ plants. If there is no behavioral preference, then the case is as described above, with selection occurring in two environments.

But suppose the distribution was not random but the result of the different behavioral preferences of the two types. The simplest case would be where $G_1$ females always (unless experimentally manipulated) deposit their eggs on $E_1$ plants, and $G_2$ females always lay their eggs on $E_2$ plants. In this case the overall fitness of $G_1$ equals the fitness of $G_1$ in $E_1$, that is, $W_0(G_1) = W(G_1,E_1)$. And $W_0(G_2) = W(G_2,E_2)$. If we let $P_i$ stand for the proportion of eggs females of genotype $i$ place in $E_1$, then the case just described is one where $P_1 = 1$ and $P_2 = 0$. Clearly, the parameters $P_i$ can range from 0 to 1. If equations (2.1) and (2.2) describe a case of habitat choice rather than unequal random distribution, then that is a case where $P_1 = 0.8$ and $P_2 = 0.2$. No matter what the values of the parameters $P_i$, the fitnesses are extrapolatable. Furthermore, although $E_1$ and $E_2$ plants are different parts of the external environment (making the external environment heterogeneous), this field is a homogeneous selective environment for these insects. Why? The relevant "phytometer" experiment here dictates that one should not start with eggs randomly distributed over the plants in

6 I will use the term "habitat choice" in a way that is meant to be neutral with respect to the mechanisms used to produce the nonrandom distribution across the external environment. Thus it does not imply any deliberation or weighing of alternatives on the part of the organism. The choice of habitat may be genetically hard wired or may result from conditioning. Alternative terms found in the literature are "habitat selection," "habitat preference," and "habitat fidelity." See Hedrick (1986) for a discussion.

the field; the unmanipulated distribution is decidedly nonrandom. Rather, one should start with mated females and allow them to deposit their eggs according to their genotype-specific preferences (their $P_i$'s). Then we would follow their offspring to sexual maturity to measure their relative (actualized) fitnesses. Recall that selective environmental heterogeneity is indicated by a genotype-environment interaction in fitness. Thus the question of whether the selective environment is spatially heterogeneous reduces to the question of whether there is an interaction between genotype and spatial position of released mated females. But that would occur only if site of oviposition depended on site of release. By hypothesis, it does not. Thus we have a single homogeneous selective environment. In this selective environment the genotype showing a preference for the richer environment has the higher relative (expected) fitness.

This may be made clearer by considering a slightly different example. Suppose as before there are two habitats, $E_1$ and $E_2$, and that selection occurs during the egg and larval stages within each habitat. More specifically, survivorship within each habitat is density dependent and each habitat contributes a fixed proportion ($c$ and $1 - c$ for habitats $E_1$ and $E_2$, respectively) to the adult mating pool. Genotypic variation is expressed only in the preferences shown by mated females for the two habitats. Thus adults have equal fecundity, and different genotypes have equal viabilities within each habitat. As before, $P_i$ stands for the proportion of eggs females of genotype $i$ leave in $E_1$. Rausher and Englander (1987) show that this system evolves to an equilibrium where there is equal survivorship in both habitats, and the overall distribution of offspring into the two habitats is in the proportion $c{:}1 - c$. Perturbation away from this equilibrium results in one of the two habitats being more densely packed than the other. Thus, since survivorship is density dependent, survivorship will decrease in the one habitat and increase in the other. Any genotype that places more offspring in the less dense habitat will have a higher fitness since its offspring will have higher than average survivorship. Such genotypes will increase in frequency until the equilibrium distribution is reached. As Rausher (1984) points out, this argument is analogous to Fisher's (1930) argument for the 1:1 sex ratio being the equilibrium sex ratio among diploid organisms. Since both sexes contribute equally to matings in diploid organisms, the rare sex is always selectively favored in that it will have a higher relative fecundity.[7]

I have argued that the random distribution of types into a heterogeneous external environment can result in a heterogeneous selective environment (it will do so if the relative fitnesses differ in different parts of the external environment), while habitat choice results in a common selective environment. This distinction is based on the sort of "phytometer" experiment that would be relevant in determining the scale of selective environmental homogeneity. In the case where organisms have no control over where they (or their offspring) live, our experimental technique should reflect this by randomly distributing the relevant genotypes through space or time. In the case of habitat choice, our "phytometer" experiment should start with that part of the life cycle where the choice is made. This reflects the fact that when habitat choice is operative, selection can act on variation in habitat preferences, as in the examples above. To treat habitat choice in the way that we have suggested that random distribution be treated would ignore or obscure this fact. On the other hand, to treat random distribution as a case of habitat choice would be to confuse "survival of the fittest" with "survival of the luckiest." This important difference is marked by differences in population genetic models; so-called multiple niche models (e.g., Felsenstein 1976 or Hedrick et al. 1976) are appropriate for cases of random distribution over a heterogeneous selective environment, whereas models of the sort discussed above (e.g., Rausher 1984 or Rausher and Englander 1987) are appropriate for cases of habitat choice.

## 2.5. RELATIONS AMONG EXTERNAL, ECOLOGICAL, AND SELECTIVE ENVIRONMENTS

Figures 2.1, 2.2, and 2.3 represent a hypothetical case of the differing scales of heterogeneity in a common field corresponding to

[7] The example presented here is a special case of a model developed by Rausher and Englander (1987), which is a generalization of a model developed by Rausher (1984). As is well known, Fisher's argument implies a 1:1 sex ratio only if the costs of producing males and females are equal. The generalized result is that there should be an equalization of parental investment in males and females at equilibrium. The result of Rausher and Englander is similarly generalized.

three different concepts of environment: external, ecological, and selective. The external environment consists of all factors external to the population of interest. Thus it consists of abiotic factors, such as nitrogen in the soil and amount of rainfall, and biotic factors, such as herbiverous insects and surrounding vegetation. The ecological envrionment reflects those aspects of the external environment that affect the target organisms' reproductive output. As illustrated by figures 2.1 and 2.2, not all external environmental variation is reflected in ecological environmental variation. One might conclude then that external environmental variation was a necessary but not sufficient condition for ecological environmental variation. However, whether or not this is true depends on whether or not we include conspecific organisms as part of the external environment. This makes a difference in cases of density-dependent population regulation. One patch in our field may be a poor spot for seedlings of, say, *Anthoxanthum odoratum* because it is already very densely packed with mature *Anthoxanthum* plants. If our concern is with the population dynamics of *Anthoxanthum*, this heterogeneity is important, but I do not think it should be considered heterogeneity of the external environment. In taking a populational (as opposed to individual) point of view, the external environment contains only factors external to the population of interest. Thus, although it may be true that most ecological environmental heterogeneity reflects external heterogeneity, such external heterogeneity is not a necessary condition for ecological environmental heterogeneity.

Heterogeneity of the selective environment reflects differing relative fitnesses in time or space (genotype-environment interaction). For the relative fitnesses of multiple types to vary in time or space, the absolute fitness (absolute reproductive output) of at least one type must vary. Thus ecological environmental heterogeneity is a necessary condition of selective environmental heterogeneity. It is not, however, a sufficient condition. Variation in the ecological environment—for instance, the spatial variation from rich to poor to rich illustrated in figure 2.2—may not show up as selective environmental heterogeneity. It will not do so if all the genotypes respond to this ecological variation in the same fashion. Such a case is illustrated in figure 2.9.

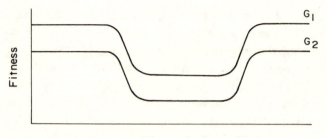

Environment

Figure 2.9. Diagram showing that variation in the ecological environment need not result in variation in the selective environment. In this case it does not because both genotypes respond to the ecological environmental variation in the same way, i.e., their relative fitness does not change.

From the point of view of the theory of natural selection, the relevant environment is the selective environment. It follows from the above considerations that external environmental heterogeneity is neither necessary nor sufficient for selective environmental heterogeneity. It is not necessary because in cases of density- or frequency-dependent selection, changes in the population structure alone can result in a change in selective environments. It is not sufficient since many changes in the external environment either do not affect the organisms at all or do not affect them differentially. And as our examples of habitat choice illustrate, external environmental heterogeneity that has the potential to produce selective environmental heterogeneity may be damped out by behavior. By behavior, organisms can reduce or eliminate the effects of some aspects of the external environment while enhancing the effects of others. Organisms can also actually modify their external environment. Thus there are many ways in which organisms buffer themselves from external environmental heterogeneity. In our discussion above about continuous environmental variation, we rejected on pragmatic and epistemological grounds the view that every organism lives in a unique selective environment. Such a view is plausible only if one equates selective and external environments. It is moderately plausible that if we measured enough external factors at a fine enough scale we could distinguish the external environments of every organism in a field. But given the many ways in which organisms can and do buffer themselves from external environmental heterogeneity, there is no reason to believe that selective environments of all species divide on such a fine grain.

These observations have considerable implications for studies of evolutionary history. For instance, if someone were to claim that the evolution of extreme phenotypic plasticity in humans was at least in part due to rapid and unpredictable environmental changes (Brandon and Hornstein 1986), the relevant environmental changes would be changes in selective environments. Such a hypothesis would thus have no necessary implications about climatic or geologic change. Conversely, the presence of major climatic or geologic changes does not imply major changes in selective pressures, that is, changes in selective environments. Unfortunately, this decoupling of selective and external environments makes the testing of certain historical hypotheses even more difficult than is generally recognized.

The distinction between external and selective environments may help resolve the dispute over the independent existence of *niches*. Richard Lewontin (1978 and 1983) argues that unoccupied niches are conceptual impossibilities, that niches have no existence independent of the organisms that occupy them. Ernst Mayr (1982) correctly points out that whether or not niches have an independent existence depends on how one defines "niche." Ecologists typically think of niches in terms of the functional roles populations play in an ecological community. A description of a niche would then be a description of the resources utilized, the parasites and predators to be avoided, the climate, and so forth. In other words, an ecological niche consists of the relevant features of the external environment. The external environment does exist independent of the population of interest. Mayr (1982) gives the following as an example: In the forests of Borneo and Sumatra there are twenty-eight species of woodpeckers. In the very similar forests of New Guinea there are no woodpeckers and no other bird species utilizing the resources available to woodpeckers.[8] Thus the

---

[8] But in New Guinea, as well as in Madagascar, which, like New Guinea, has never been colonized by woodpeckers, specialized mammals do utilize the woodpecker niche. In Madagascar the primate *Daubentonia madagascariensis* fills this

woodpecker niche exists in New Guinea. But Lewontin's concept of environment is clearly different. He is concerned with the environment as it determines selective pressures, as it determines the process of adaptation. He is therefore concerned with the selective environment. As we have seen above, selective environments have no existence independent of the population of organisms of interest. Figure 2.6 illustrates how the pattern of selective environmental heterogeneity depends on the presence or absence of certain genotypes. Note that this example does not involve density- or frequency-dependent selection. But density- and frequency-dependent selection, as well as habitat choice and other behavioral methods of buffering external environmental heterogeneity, provide cases where the nature of the selective environment depends on the relevant organisms. Although I think the term "niche" should be reserved for its ecological usage, if niches are selective environments then Lewontin is right in saying that unoccupied niches are conceptual impossibilities.

This dispute is more than a semantic quibble. Lewontin's (1978 and 1983) major point is that the metaphor of the environment independently posing problems and the organisms evolving by natural selection to solve these problems is inappropriate for the process of evolution by natural selection. According to this metaphor the environment is like a lock, and natural selection shapes a key to fit it. But if the organisms are constitutive of the selective environment, if they in a way "construct" their own environment, this metaphor is clearly misleading. To many, Lewontin's critique reflects a defeatist attitude. Indeed, if our only methods of determining patterns of environmental heterogeneity depended on measuring features of the external environment—features independent of the organisms of interest—and if we wanted to determine patterns of selective environmental heterogeneity, then our task would be impossible. But the methods recommended above necessarily involve the organisms of interest: they are our measuring "instruments." The dependence of the environment on the organisms experiencing selection poses neither conceptual nor methodological problems for our notion of selective environments.

niche, while in New Guinea it is filled by marsupials of the genus *Dactylopsila*. See Cartmill (1974).

In this section I have argued that external and selective environments are decoupled because heterogeneity in one is neither a necessary nor a sufficient condition for heterogeneity in the other. This does not mean that the external environment is irrelevant to evolution by natural selection. Fish may have fins, but also mammals (e.g., whales and seals) and birds (e.g., penguins) that spend much of their life swimming have developed finlike appendages—clear evidence that water is a part of all of these animals' selective environments and that they have adapted to it. Striking cases of mimicry such as the Viceroy butterfly's resemblance to distasteful Monarchs also are good presumptive evidence of the real existence of predators. Thus both abiotic (e.g., water) and biotic (e.g., predators) features of that part of the environment external to the evolving population can drive evolution by natural selection. The point is that environmental features internal to the evolving population may also drive that process. But at present we do not have sufficient evidence to generalize on the relative importance of these two sorts of environmental factors in evolution.

## 2.6. Common Causes and Common Environments

A selective environment is an area (or population; recall figure 2.5b) that is homogeneous with respect to the relative fitness of a set of competing types. (In natural populations the relevant set of types is the set of types actually extant in the population.) Elliott Sober's (1984) use of the notion of common cause may lead one to think that homogeneity of fitness values defines a *type* of selective environment of which there could be indefinitely many *tokens*. (For example, the U.S. penny is a type of coin of which any particular penny is a token.) According to Sober (1984, p. 280) group selection requires that there be a common causal influence that affects the fitness of each member of the group equally. Similarly, one might think that two organisms are in a common selective environment only if some common cause, some common selective force, affects both of them. For instance, suppose there are two geographically separate populations of gazelles in which there is directional selection of equal intensity for increased speed, making both populations homogeneous with respect to the relative fitnesses of the competing types. But one population is preyed on by cheetahs, the other by lions. Thus members of the two populations are not affected by a common cause or a common selective force even though the effects are the same. And so, one might conclude, the two populations do not share a common environment but inhabit different tokens of the same type of environment.

Although this distinction sounds attractive and simple enough, it raises serious, perhaps insuperable, conceptual problems. A single cheetah is a common cause of the death of two gazelles if it kills both of them. But within a population do two gazelles eaten by two different cheetahs experience a common cause of death? If so, how about two gazelles within a population, one killed by a cheetah, the other by a pride of lions? We have two options. First, we might say that these two gazelles experience different selective pressures. Then it follows that a single population of gazelles that experiences predation by both lions and cheetahs is divided into at least two parts corresponding to two different selective environments. What then do we say about a gazelle that escapes both lions and cheetahs? Does it inhabit both selective environments, or is there a third selective environment corresponding to the selective pressure resulting from the predation of both lions and cheetahs? Second, we could say that our two gazelles experience a common selective pressure. But how could we justify saying that they inhabit a common selective environment when we concluded above that the gazelles in the two geographically separate populations inhabit different tokens of the same type of selective environment?

If we pursue such questions in a systematic and rigorous way, we will find that the distinction of different tokens of the same type of selective environment based on differences in the causes of relative fitness is really quite muddled. Moreover, even if the distinction could be drawn in a rigorous way, its utility would be limited. Insofar as we are concerned with current evolutionary dynamics of a given population, the causes of differential reproduction are irrelevant.[9] They are also very difficult to determine.

[9] This is not to say that the causes of differential reproduction are irrelevant to all evolutionary questions. As we have already seen in chapter 1, we need this information to fully instantiate the PNS. See the further discussion in chapter 5,

Suppose we discover the patterns of selective environmental heterogeneity in a field by means of a "phytometer" experiment. We could then try to correlate this pattern with some feature or cluster of features of the external environment. Such a task is laborious and may not yield any interesting results (see Antonovics, Ellstrand, and Brandon 1988 for some of the pitfalls of trying to do this). Even if we do find an interesting correlation, we may not know whether the correlation is a causal relation or a common effect of some as yet unmeasured causal factor. All this effort yields precious little if we are interested in understanding the evolutionary dynamics of the target population.

We can, of course, distinguish different tokens of the same type of selective environment in terms of spatiotemporal separation. Thus we could say that the two geographically separate populations of gazelles discussed above do inhabit two different tokens of the same type of selective environment. But the basis of the distinction would be spatiotemporal separation, not the lack of a common cause that impinges on both populations.

The basis of my criticism of the idea that the organisms inhabiting a particular selective environment must all be affected by a common cause is not a lack of operationality. I have criticized the idea because it has limited utility. For our purposes, selective environments can be characterized in terms of the patterns of relative fitnesses without regard to the sources of that pattern.

## 2.7. Genealogical Groupings, Ecological Groupings, and the Consequences of Selection in Heterogeneous Environments

The usual way of dividing up the biological world is based on the flow of genes through time and space. The species category is fundamental in this categorization. According to the biological species concept, species are defined in terms of interbreeding; they are the largest groups within which interbreeding occurs (see Mayr 1987). In sexual species patterns of interbreeding determine the flow of genes through space. According to the phylogenetic

---

where it will become clear that in answering what-for questions, questions concerning the evolution of adaptations, such information becomes relevant.

species concept, species are defined in terms of monophyletic genealogy (see Mishler and Brandon 1987). This can be thought of as gene flow through time. As different as these conceptions are, they are both fundamentally based on gene flow. Of course there are higher level genealogical units, for example, *clades*. Clades are monophyletic taxa consisting of a chosen species and all its descendant species. Species themselves can be subdivided on the basis of gene flow. A *deme* is a population within which breeding is random. Some species may be divisible into discrete demes, while others, so-called viscous species, are divisible into *genetic neighborhoods* (S. Wright 1946). Again these categories are directly based on gene flow. The hierarchy of demes or genetic neighborhoods, species, and clades may be thought of as part of a genealogical hierarchy (Eldredge 1985).

In recent years a number of biologists have recognized that the entities picked out by the process of gene flow need not correspond to the entities picked out by other biological processes, in particular ecological processes (Wilson 1980; Mishler and Donoghue 1982; Damuth 1985; Eldredge 1985). This recognition corrects some longstanding and erroneous assumptions. For instance, many have simply assumed that species and demes are ecologically and/or selectively homogeneous. Van Valen (1971, 1976) even incorporated this assumption in his definition of species. It is probably not much of an exaggeration to say that virtually every empirical experiment that has had the power to test this assumption has refuted it. The relative fitnesses of varying types within a species or local population vary through time and space; genotype-environment interactions are there to be found if one looks for them (Sultan 1987).

D. S. Wilson (1980) recognized that demes need not be selectively homogeneous. His concern was with frequency-dependent selection where the frequency-dependent effects are felt within small groups and the frequency of a type in one group has no effect on other groups. Such groups are called *trait groups* and they are selectively homogeneous with respect to some particular trait. Wilson makes the point that different traits can pick out different trait groups. For instance, in some insects interactions during the larval stage may take place among a relatively small group, say the group on some host plant. If there is frequency-dependent selection for some larval trait, the relevant frequency is the frequency of the trait among the larvae on that plant. Those larvae form a trait group with respect to that trait. If adults disperse into a much larger population for mating, then with respect to a trait affecting mating success the trait group would be this larger population. We have described selective homogeneity in terms of overall fitness; but if we only measured some one fitness-affecting trait, such as a trait affecting larval survival or adult mating success, then our selectively homogeneous populations would be populations that are selectively homogeneous with respect to that trait and thus would be Wilsonian trait groups. Wilson (1980) argues that in general trait groups are smaller than demes. Whether or not this is correct, it raises the question of the relative sizes of selectively homogeneous populations and demes (or genetic neighborhoods). Of course, this is an empirical question and it is safe to say that it will receive different answers for different populations. The generalization we can draw is that there is no reason to expect a correspondence between demes (or genetic neighborhoods) and selectively homogeneous populations, that is, we cannot assume that demes (and certainly not species) are selectively homogeneous. The failure of this assumption is relevant to some of the central questions of contemporary evolutionary biology.

I will discuss three areas where selection in heterogeneous environments is invoked to explain important and pervasive biological phenomena. Before doing so let me eliminate one source of potential confusion. I have argued that natural selection requires common selective environments, that is, selection occurs within selective environments but the process of differential reproduction occurring across selective environments (e.g., the example discussed in section 2.4) is not simply natural selection. It is a process consisting of natural selection within environments and distribution into environments. That of course does not mean that we do not want models to describe this more complex process. (The multiple-niche models mentioned in section 2.4 are of this type.) Following Damuth (1985) I will call selection within a selectively homogeneous environment *simple natural selection*, and will introduce the term *compound natural selection* to describe this more complex process of distribution and selection across selectively heterogeneous environments.

The first area where compound selection has been invoked to explain a pervasive biological phenomenon concerns the prevalence of sexual reproduction in the biosphere. Sexual reproduction seems problematic from a genetic point of view because there is a 50% cost to meiosis. By sexually reproducing, a female reduces her genetic contribution to her offspring by 50%. In light of this apparent disadvantage of sex, theoreticians have offered a variety of models to account for the advantage of sex and genetic recombination (see Ghiselin 1974; G. C. Williams 1975; Maynard Smith 1978; Bell 1982; Michod and Levin 1988). Some of these models are group selection models; they seek to find a "long-term" advantage to sex. The basic idea behind them is that sexually reproducing groups will be more evolutionarily responsive to environmental changes than will asexual groups. Thus by a process of differential extinction sexual reproduction will become more frequent. (In chapter 3 we will discuss some of the problems faced by such group selection explanations.) Another class of models seeks "short-term" advantages to sex. Among these, some argue that the production of variable progeny is favored when environments vary in time and space. Consider the case of temporal variation. If environments change from generation to generation and become negatively autocorrelated, then the production of progeny that are different from the parents will be favored, and thus sexual reproduction will be favored. Similarly, if the spatial scale of environmental heterogeneity is such that dispersed offspring are likely to be in environments different from that of the parents, then again sex will be favored.

Although we may think of the environmental heterogeneity posited in these models as heterogeneity of the external environment, when we explicitly translate these models into the fitness effects of different genotypes in different environments it becomes clear that the selective environment is the relevant one. Indeed, an extreme type of selective heterogeneity is required for the models to work. The ordinal relation in the fitnesses of the different genotypes must change across environments, that is, the genotype-environment interaction must be of the "crossing type" (see Antonovics, Ellstrand, and Brandon 1988), which is graphically represented by the crossing of the fitness lines of different genotypes across an environmental gradient (as in figure 2.7). This makes intuitive sense. If the ordinal relation of the fitnesses of the different genotypes did not change across space or time, then an asexual reproducer of the right genotype, that is, the genotype that is everywhere and at every time the fittest, would always be selectively favored over sexual reproducers of variable progeny.

The second area where compound selection has been invoked to explain pervasive biological phenomena concerns the amount of genetic (and phenotypic) polymorphism found in natural populations. New molecular techniques for detecting genetic variation have detected a surprising amount of variation. Such diversity is surprising because if selection were acting in a uniform way throughout the population one would not expect so many variants to be present at such high frequencies. This evidence has led to the selectionist-neutralist debate (Lewontin 1974). The neutralists maintain that much of this genetic variation is selectively neutral, and that the variation found in natural populations is due to mutations and genetic drift. Although the amount of observed variation could not be explained by selection if demes and local populations were selectively homogeneous, if they are selectively heterogeneous selection might be able to maintain such levels of variation. The basic idea is simple: if different genotypes are favored at different times or in different regions, then genetic polymorphisms can be maintained by selection. Hedrick (1986) has reviewed the models seeking to offer such explanations. As with models showing a short-term advantage of sex, these models require a crossing type genotype-environment interaction. For instance, in the model that offers the least stringent conditions under which selection may maintain a genetic polymorphism, Gillespie's (1976) SAS-CFF model, the environments are formally characterized in terms of the relative fitnesses in the two environments. The model is for a single-locus with two alleles. Environment 1 is characterized by the fitness values, 1, $1 - ks$ and $1 - s$ for genotypes $AA$, $Aa$, and $aa$, respectively, while in environment 2 the fitness values of the homozygotes are reversed. $1 - s$, $1 - ks$, and 1 are the fitness values of $AA$, $Aa$, and $aa$, respectively.[10]

---

[10] $s$ is the selection coefficient and $k$ stands for the level of dominance. $k$ will

Again it is clear that it is the selective environment that is relevant here.

There is one final area in which the nature of selection across heterogeneous environments is of potential relevance to the explanation of important biological phenomena. In the first two instances compound selection is invoked to explain observations that are seemingly problematic for the theory of evolution by natural selection (sex and high levels of genetic variation within populations). In this final area there are no definitive observations, but there is an important question: At what scale do we expect to find genetic differentiation, that is, at what scale do we expect to find adaptation to local environments? To answer this question, which has received some theoretical attention (see Slatkin 1973 and Roughgarden 1979, chap. 12), we must consider the relative sizes of demes and selectively homogeneous environments. Theoreticians have found that whether or not a population can adapt to environmental heterogeneity depends on the scale of selective environments, the scale of gene flow, and the process of population regulation (the scale of the ecological environment). Although quite complicated, the basic idea makes sense. Genes selected in different environments mix during mating and thus slow down or even stop the adaptation to local conditions. Thus, everything else being equal, one would expect species that have demes (or genetic neighborhoods) smaller than or equal to their selective environments (or selective neighborhoods) would show greater micro-adaptation (and hence greater genetic differentiation) than species where that relative size is reversed. This hypothesis has received little, if any, experimental attention.[11]

---

always be a positive number less than 1. Note that this implies a reversal of dominance between the two environments, that is, the heterozygote is always closer in fitness to the favored homozygote.

[11] Janis Antonovics and I are currently testing this hypothesis. We are comparing two grasses, *Anthoxanthum odoratum* and *Danthonia spicata*, in a field where both occur naturally. In one half of the experiment we are measuring the spatial scale of selective environmental heterogeneity; in the other half we are measuring the amount of microdifferentiation. The two species were chosen since they do differ in the size of their genetic neighborhoods. Studies of the evolution of heavy-metal tolerance in plants (Antonovics, Bradshaw, and Turner 1971) have revealed some evidence of the evolutionary adjustment of the size of genetic neighborhoods in response to radical changes in selective environments. See further discussion in chapter 5.

Once we recognize that selectively homogeneous populations need not correspond to any genealogical unit, we can formulate interesting evolutionary questions. To what extent do organisms have control over the size of their genetic neighborhoods or the size of their selective neighborhoods? From an evolutionary point of view, what is the optimal relative size of the two? As pointed out above, if selective neighborhoods are smaller than genetic neighborhoods, adaptation to local environments is slowed if not prevented. Should we expect evolutionary adjustment of either selective or genetic neighborhoods to prevent this? Perhaps; but there may be evolutionary costs to micro-adaptation at too fine a scale. Perhaps instead we should expect the evolution of phenotypically plastic genotypes, which do fairly well across a range of selective environments (Bradshaw 1965; Sultan 1987).

Whatever the answer to these questions, the major point here is that whether one is interested in the evolutionary advantage of sex, the possibility of selection maintaining genetic polymorphisms, the scale of micro-adapation, or the advantages of phenotypic plasticity, the relevant notion of environment is that of the selective environment. This is implicit in the models mentioned above in that different environments are formally represented by sets of fitness values for the types; a difference in relative fitnesses means a difference in environment. To understand the possible applications of models of selection in heterogeneous environments, we need to understand what a selective environment is; to test these models we need to be able to put this conception into operation. My aim in this chapter has been to provide the relevant concept and the methods for its operationalization.

CHAPTER 3

# *The Levels of Selection*

Biologists have long recognized that the biosphere is hierarchically arranged. And at least since 1970 we have recognized that the abstract theory of evolution by natural selection can be applied to a number of elements within the biological hierarchy (Lewontin 1970). But what is it for selection to occur at a given level of biological organization? What is a "unit of selection"? Is there one privileged level at which selection always, or almost always, occurs? In this chapter I will try to clarify and partially answer these questions. In the first four sections I explicate the distinction between replicators and interactors, a generalization of the genotype-phenotype distinction, and demonstrate how this distinction is relevant to questions concerning the level or levels at which selection occurs. I present a hierarchy of interactors, that is, a hierarchy of levels of selection, and a derivative "hierarchy" of replicators. The next four sections are devoted to some questions concerning group selection and its relation to organismic selection. In the final section I explore the space of alternative approaches.

## 3. 1. INTERACTORS AND REPLICATORS

In Chapter 1 I offered a simple description of the process of evolution by natural selection. Recall that the process consists of three steps. The first step is the selective discrimination of phenotypes. For example, in the case of directional selection for increased height, the first step is the differential reproduction of organisms of differing heights. The form differential reproduction takes can vary (e.g., differential survivorship, differential fecundity), and the ecological reasons for the greater success of taller organisms can vary. In any case, taller organisms tend to have greater reproductive success than shorter ones. Selection can be thought of as an interaction between phenotype and environment that results in differential reproduction. It requires phenotypic variation. But natural selection in this sense (what quantitative geneticists call "phenotypic selection") is not sufficient to produce evolutionary change. In the case of directional selection for increased height, selection may change the phenotypic distribution in the parental generation (it will do so if selection is by differential mortality); but whether or not that results in evolutionary changes, that is, in changes in the next generation, depends on the *heritability* of height. That is, it depends on whether or not taller-than-average parents tend to produce taller-than-average offspring and shorter-than-average parents tend to produce shorter-than-average offspring. This is the second step in the process. Of course, height is not directly transmitted from parent to offspring; rather, genes are.[1] Thus offspring of taller-than-average parents will tend to have genotypes different from those of offspring of shorter-than-average parents. To go full circle, that is, to get to an evolutionary change in the phenotypic distribution, these genotypes must *develop*. By the process of epigenesis (the development of a zygote into the mature phenotype), these genotypic differences manifest themselves as phenotypic differences. And so the phenotypic distribution of the offspring generation has been altered; evolution by natural selection has occurred.

Thus evolution by natural selection requires both phenotypic variation and the underlying genetic variation. In one step, phenotypes interact with their environment in a way that causes differential reproduction. This leads to the next step, the differential replication of genes. Through epigenesis, this new genotypic distribution leads to a new phenotypic distribution and the process starts anew. Development, or epigenesis, is the link between genotype and phenotype. Its importance will be discussed later in this chapter, but for now we will concentrate on the first two steps of the process.

The above description of evolution by natural selection seems perfectly adequate for cases of selection occurring at the level of organismic phenotypes. But during the last twenty-five years there has been increasing interest in the idea that selection may

[1] Nuclear genes are not the only means of transmitting traits from parent to offspring. Among other means, cytoplasmic DNA and culture are prominent.

occur at other levels of biological organization. This interest was sparked by V. C. Wynne-Edwards's book, *Animal Dispersion in Relation to Social Behavior* (1962). Wynne-Edwards argued that a major biological phenomenon, the regulation of population size and density, evolves by *group selection*. In reaction to this thesis, G. C. Williams (1966) and, later, Richard Dawkins (1976) argued that selection occurs primarily at the level of genes. In recent years there has been a flurry of theoretical investigations into kin and group selection with some explicitly hierarchical models resulting.[2] It is not obvious how we should apply the genotype-phenotype distinction to describe cases of gametic selection, group selection or species selection. Thus David Hull (1980 and 1981) and Dawkins (1982a and 1982b) have introduced a distinction between *replicators* and *interactors* that is best seen as a generalization of the traditional genotype-phenotype distinction.[3]

Dawkins defines a replicator as "anything in the universe of which copies are made" (1982a, p. 83). Genes are paradigm examples of replicators, but this definition does not preclude other entities from being replicators. For instance, in asexual organisms the entire genome would be a replicator; and in cultural evolution, ideas—or what Dawkins calls *memes*—may be replicators (Dawkins 1976).

The qualities making for good replicators are longevity, fecundity, and fidelity (Dawkins 1978). Here longevity means longevity in the form of copies. It is highly unlikely (indeed in most cases impossible) that any particular DNA molecule will live longer than the organism in which it is housed. What is of evolutionary importance is that it produce copies of itself so that it is potentially immortal. Of course, everything else being equal, the more copies a replicator produces (fecundity) and the more accurately it produces them (fidelity), the greater its longevity and evolutionary success.

---

[2] See Brandon and Burian (1984) for a collection of some of the more important papers concerning the levels of selection. The chapters by Hamilton, Wimsatt, and Arnold and Fristrup offer hierarchical models of selection.

[3] More precisely, this distinction generalizes only a part of the traditional genotype-phenotype distinction, that part which relates to selection. As we will see below, there are terminological as well as substantive differences between Hull and Dawkins.

In explicating Dawkins's notion of replicators, Hull stresses the importance of directness of replication. Although according to Dawkins organisms are not replicators, they may be said to produce copies of themselves. This replication process may not be as accurate as that of DNA replication, but nonetheless there is a commonality of structure produced through descent from parent to offspring. However, there is an important difference in the directness of replication between these two processes. The height of a parent is not directly transmitted to its offspring; as discussed earlier in this section, such transmission proceeds indirectly through genic transmission and ontogeny. Genes, however, replicate themselves less circuitously. Both germ-line replication (meiosis) and soma-line replication (mitosis) are physically quite direct. The importance of this is made explicit by Hull, who defines replicators as "entities which pass on their structure directly in replication" (1981, p. 33).

As I stated earlier, evolution by natural selection involves two steps (we will ignore development for now). One step involves the direct replication of structure, and the other involves some interaction with the environment so that replication is differential. The entities functioning in the latter step have traditionally been called phenotypes. But if we want to allow that biological entities other than organisms can interact with their environment in ways that lead to differential replication, then we need to generalize the notion of phenotype. To this end Hull (1980, 1981) suggests the term "interactor" and defines it as "an entity that directly interacts as a cohesive whole with its environment in such a way that replication is differential" (1980, p. 318).

Although Hull and Dawkins largely agree on the replicator-interactor distinction, two differences are worth noting. The first is purely terminological. Dawkins has not adopted Hull's term "interactor"; instead he uses "vehicle." According to Dawkins, a vehicle is "any relatively discrete entity, such as an individual organism, which houses replicators . . . and which can be regarded as a machine programmed to preserve and propagate the replicators that ride inside it" (1982b, p. 295). Because I prefer Hull's term and definition, I will use "interactor."

The second difference is more substantive. Dawkins holds that any change in replicator structure is passed on in the process of replication (1982a, p. 51, 1982b, p. 85). Thus, given the truth of Weismannism (the doctrine stating that there is a one-way causal influence from germ line to body), replicators are supposedly different from most interactors (e.g., organisms). But DNA is capable of self-repair, and so not all changes in DNA structure are passed on in the process of replication. Thus the property of transmitting changes in structure during replication does not sharply demarcate replicators from interactors.[4] What seems to be important is that replication be direct and accurate. But both "directness" and "accuracy" are terms of degree, and if we allow some play in both, then under certain circumstances an organism could be a replicator (which would not preclude its being an interactor as well). For example, Hull (1981, p. 34) argues that a paramecium dividing into two can be considered a replicator since its structure is transmitted in a relatively direct and accurate manner.

It is important to note that the definitions of interactors and replicators are given in functional terms, that is, in terms of the roles these entities play in the process of evolution by natural selection. Nothing in the definitions precludes one and the same entity from being both an interactor and a replicator. For instance, it is likely that the self-replicating entities involved in the earliest evolution of life on this planet were both interactors and replicators (see Eigen et al. 1981). Likewise, in cases of meiotic drive parts of chromosomes, or perhaps even entire chromosomes, can be considered interactors as well as replicators. As Hull suggests, in some cases organisms can be considered replicators as well as interactors.

## 3.2. LEVELS OF SELECTION

Having developed the notions of replicator and interactor, generalizations of notions of genotype and phenotype, we may now ask whether the process of evolution by natural selection occurs at levels of biological organization other than the organismic. This

[4] Besides, as Buss (1983) has shown, Weismann's separation of the germ line from the soma does not generally hold. See the discussion in section 3.3.

seemingly simple question is actually ambiguous.[5] Are we asking whether or not there are replicators other than single genes, or are we asking whether there are interactors other than organismic phenotypes? Both are interesting questions, but unfortunately many who have addressed *the* units of selection question have failed to notice that there are two separate questions and thus have confused the two. I will discuss this problem in the next section, but here I will concern myself with interactors.

Why are standard cases of selection organismic?[6] Put another way, what features of standard cases of organismic selection make organisms the interactors? What justifies our claim that in such cases "natural selection favors (or discriminates against) phenotypes, not genes or genotypes" (Mayr 1963, p. 184)? Consider again our example of directional selection for increased height. Recall that taller organisms have a higher fitness on average than shorter organisms; thus there is a positive association between height and fitness. But there is genetic variation in height, so there is also a positive association between certain genes and/or genotypes and fitness. Why not say then that natural selection favors phenotypes and genes (or genotypes) equally? Where is the asymmetry between phenotype and genotype?

The asymmetry is this: reproductive success is determined by phenotype irrespective of genotype. At an intuitive level, selection "sees" a 4-foot plant as a 4-foot plant, not as a 4-foot plant with genotype $g$. This idea can be made precise by use of the probabilistically defined notion of *screening off* (Salmon 1971). The basic idea behind the notion of screening off is this: if $A$ renders $B$ statistically irrelevant with respect to outcome $E$ but not vice versa, then $A$ is a better causal explainer of $E$ than is $B$. In symbols, $A$ screens off $B$ from $E$ if and only if $P(E,A \cdot B) = P(E,A) \neq P(E,B)$ [read '$P(E,A \cdot B)$' as the probability of $E$ given $A$ and $B$]. If $A$ screens off $B$ from $E$ then in the presence of $A$, $B$ is statistically irrelevant to $E$, that is, $P(E,A) = P(E,A \cdot B)$. But note that this relation between $A$ and $B$ is not symmetric. Given $B$, $A$ is still statistically

---

[5] I believe that Hull (1981), Dawkins (1982a), and Brandon (1982) arrived at this conclusion independently.

[6] In this context I prefer the term "organismic selection" to the more common "individual selection" because, as Hull has pointed out, interactors at other levels (e.g., groups) must be individuals.

relevant to $E$, that is, $P(E,B) \neq P(E,A \cdot B)$. Thus, where $A$ and $B$ are causally relevant to $E$, it follows that $A$'s effect on the probability of $E$ acts irrespective of the presence of $B$, but the same cannot be said of $B$. The effect $B$ has on the probability of $E$ depends on the presence or absence of $A$. For our purposes the important point is that proximate causes screen off remote causes from their effects.[7]

Let us return to our case of directional selection for increased height. In this case there is differential reproduction of interactors (organisms) and replicators (genes). But it is obvious that the means by which genes replicate differentially here is the differential reproduction of organisms. (In other words, there would be no differential replication of genes without the differential reproduction of organisms.) So why do taller organisms tend to leave more offspring than shorter organisms? Using the notion of screening off, we can see that this is best explained in terms of differences in height rather than in genotype.

We need to show that for any level of reproductive success $n$, phenotype $p$, genotype $g$, and selective environment $s$, $P(n,p \cdot g \cdot s)$ = $P(n,p \cdot s) \neq P(n,g \cdot s)$. *Gedanken* experiments should suffice to show the correctness of both the equality and the inequality. Basically, manipulating the phenotype without changing the genotype can affect reproductive success (castration is the most obvious example). On the other hand, tampering with the genotype without changing any aspect of the phenotype cannot affect reproductive success. Admittedly, the latter claim is not straightforwardly empirical. One could tamper with germ-line DNA, say by irradiation, and negatively affect reproductive success without *obviously* affecting the phenotype, but I would argue that in every such case one could find some aspect of an interactor that had been affected. For example, in many cases of irradiation of a male, sperm morphology and behavior are changed. The claim is that a change in the informational content of the genome alone will not make for a change in reproductive success.[8] Thus the fact that phenotype

---

[7] For clarity of exposition I have not explicitly relativized these probabilities to some set of conditions, but we should, as we will later in this chapter.

[8] As we saw earlier, the notions of interactor and replicator are not mutually exclusive; one and the same entity can be both interactor and replicator. Similarly, the notions of genotype and phenotype are not mutually exclusive. The genotype of an organism is a part of its phenotype. Thus my claim commits me to the posi-

screens off genotype from reproductive success shows that there is an asymmetry between phenotype and genotype, and that in cases of organismic selection reproductive success is best explained in terms of properties of the phenotype. What is true of this relation between phenotype and genotype obviously holds for the relation between phenotype and gene.

It may seem that this conclusion is a product of choosing to look at differential reproduction of interactors (organisms) rather than replicators.[9] In our case taller organisms outreproduce shorter organisms, and it should be clear that this is the mechanism by which some genes outreproduce others. But let us change our focus. Let $n$ stand for the realized fitness of a given germ-line gene, let $p$ stand for the phenotype of the organism in which it is housed, and let $g$ stand for some property of the gene (its selection coefficient or whatever else we might think is relevant). Still, the phenotype of the organism screens off the genic property from its own reproductive success; that is, $P(n,p{\cdot}g{\cdot}s) = P(n,p{\cdot}s) \neq P(n,g{\cdot}s)$. Thus in our case a particular gene's reproductive success is best explained in terms of the height of the organism in which it is housed.

I have argued that in standard cases of organismic selection the mechanism of selection, or the differential reproduction of organisms, is best explained in terms of differences in organismic phenotypes, because phenotypes screen off both genotypes and genes from the reproductive success of organisms. Thus in such cases the interaction between interactor and environment that leads to

---

tion that any change in genotype that does lead to a change in reproductive success must also be a change in the organism's phenotype. This position should not be seen as counterintuitive so long as we remember that genes (lengths of DNA) have a physical structure.

[9] Sober (1984, pp. 229–230) has raised this objection. He writes: "Brandon chose an organism's reproductive success. But suppose we choose *change in gene frequencies*. Then the screening-off relation is inverted. Gene frequencies and genic selection coefficients *determine* change in gene frequencies, if the population is infinitely large, and confer a probability distribution on future gene frequencies, if drift is taken into account." This objection is based on a simple equivocation. In the first instance, we are concerned with relations among objective probabilities in the real world. That is the sense in which height, not genotype, determines reproductive success. Sober is concerned with the relation between coefficients and variable values in a mathematical model. Mathematical determination in a model does not translate so simply to nature.

differential reproduction occurs at the level of the organismic phenotype.

We can now return to the question with which we began this section: Do such interactions occur at other levels of biological organization? Are there other levels of interactors besides those of the organismic phenotype? This is ultimately an empirical question and I do not intend to answer it definitively; rather, I shall try to offer the conceptual tools necessary to answer it. But let us first consider selection at the group level before defining levels of selection.

Group selection is natural selection acting at the level of biological groups. And natural selection is the differential reproduction of biological entities that is *due to* the differential adaptedness of those entities to a common environment. I have already defended this definition in chapter 1 (also see Brandon 1978 and 1981b), but two points are worth reemphasizing. First, the definition is explicitly causal; thus it does not include all cases of differential reproduction. For instance, it does not apply to cases where by chance a less well adapted organism has greater reproductive success than a better adapted one (which, let us say, was struck by lightning). Second, it applies only to those cases where differences in reproductive success are due to differences in adaptedness to a *common selective environment*. This is implicit in the above discussion where I moved from saying that the organism's phenotype best explains its level of reproductive success, to saying that differences in phenotypes best explain differences in reproductive success. This move is valid only if we restrict our attention to organisms, or more generally interactors, within a common selective environment. This is illustrated by the simple example discussed in chapter 2. Suppose we plant two seeds, one in good soil, the other in mildly toxic soil. The first will probably survive longer and produce more seeds than the second, that is, the first will be "fitter" than the second. But to explain this difference we must refer to differences in their environments, not differences in their phenotypes.

As we have seen above, organismic selection is the differential reproduction of organisms that is due to the differential adaptedness of those interactors to a common environment. By analogy, then, group selection would be the differential reproduction of biological groups that is due to the differential adaptedness of those groups to a common environment. Thus a necessary condition for the occurrence of group selection is that there be differential reproduction (propagation) among groups.[10] But this necessary condition is not sufficient. For the differential reproduction of groups to be group selection, that is, selection at the group level, there must be some group property (the group "phenotype") that screens off all other properties from group reproductive success.[11]

It is by no means necessary that such a property exist. For instance, suppose that group productivity or group fitness depends simply on the number of organisms within the group at the end of a certain time period.[12] Suppose further that the adaptedness values of these organisms do not depend in any way on the group composition. In that case the group "phenotype" (the distribution of individual phenotypes within the group) would not screen off all nongroup properties from group reproductive success. In particular, it would not screen off the aggregate of the individuals' phenotypes, that is, the following relation would *not* hold: $P(n, G \cdot [p_1 \cdot p_2 \cdot \ldots \cdot p_k] \cdot s) = P(n, G \cdot s) \neq P(n, [p_1 \cdot p_2 \cdot \ldots \cdot p_k] \cdot s)$, (where $n$ is the number of propagule groups, $G$ the group phenotype, and $p_i$ the phenotype of the $i^{\text{th}}$ member of the group). The equality would hold, but the inequality would not, since the phenotype of each individual within the group would determine that individual's adaptedness, and the adaptedness values of each member of the group would determine the adaptedness value of the group.

In summary, group selection occurs if and only if (1) there is

---

[10] This claim will be discussed in detail in section 3.6.

[11] It is not completely clear what should count as group properties. Obvious examples include properties that could not belong to individual organisms, such as the relative frequency of certain alleles within the group, the phenotypic distribution within the group, or the geographic distribution of the group. Other properties that might be selectively relevant are less obviously group properties. For instance, we may or may not want to count the ability to avoid predation as a group property. Whether or not it is would depend on whether the group's ability to avoid predation is something "over and above" the ability of each individual to avoid predation, that is, whether there is some group effect on the individuals' abilities to avoid predation.

[12] This need not be the case, but is assumed in most models. For a review of these models, see Wade (1978), Uyenoyama and Feldman (1980), Wilson (1983a), or the introduction to part 3 in Brandon and Burian (1984). Indeed, in one experimental treatment, Wade (1977) selected for groups with the lowest numbers of organisms.

differential reproduction of groups; and (2) the group phenotype screens off all other properties (of entities at any level) from group reproductive success. One way to restate clause (2) is this: Differential group reproduction is *best* explained in terms of differences in group-level properties (differences in group adaptedness to a common selective environment). Still another way to say the same thing—a way that would have seemed question-begging before we discussed screening off—is this: The causal process of interaction occurs at the level of the group phenotype. (Note that clause (2) implies that the differential reproduction mentioned in clause (1) is nonrandom and is due to differential adaptedness.)

What has been said about group selection can be easily generalized into the following definition: Selection occurs at a given level (within a common selective environment) if and only if

1. There is differential reproduction among the entities at that level; and
2. The "phenotypes" of the entities at that level screen off properties of entities at every other level from reproductive values at the given level.

### 3.3. A HIERARCHY OF INTERACTORS

What sorts of biological entities fall under the definition given above? Organisms certainly do (for ample documentation, see Endler 1986). What about entities at lower levels of biological organization? Eigen and co-workers (1981) have presented a plausible scenario concerning the origin of life. In this scenario lengths of RNA interact with proteins in a "primordial soup," and by this selection process the genetic code develops. Thus, in this scenario, there is selection at the level of lengths of RNA. Clearly these bits of RNA qualify as replicators; they replicate their structure directly and accurately. But they are interactors as well. It is their physical structure, their "phenotype," that determines their adaptedness to given conditions. For instance, in one experimental treatment RNAs were selected under conditions of high concentrations of ribonuclease, an enzyme that cleaves RNA into pieces. In this treatment variants that were resistant to cleavage evolved. "Apparently the variant that is resistant to this degradation folds in a way that protects the sites at which cleavage would take place." (Eigen et al. 1981, p. 97).

Doolittle and Sapienza (1980) and Orgel and Crick (1980) have argued that intragenomic selection results in the spread of "selfish genes," that is, genes that increase their representation in the genome not through their effects on the phenotype of the organisms in which they are housed, but through their superior replication efficiency within the genome. Such genes may or may not be transcribed, but in general one expects them to have a negative impact on the fitness of organisms because of the energetic costs of excess DNA. Doolittle and Sapienza (1980) describe the selection process by which "selfish genes" spread as "non-phenotypic selection." In the terminology of this chapter, what they mean is that the level of this selection process is not the organismic phenotype. But "selfish genes" are interactors. They interact within the cellular environment in a way that leads to differential replication. It is their "phenotype" (i.e., their physical structure) that matters, not the phenotype of the organism in which they are housed. Similar remarks apply to chromosomes or parts of chromosomes in cases of meiotic drive (see Crow 1979).

Genic selection in the sense described above and meiotic drive are selection processes that occur within a cellular environment. One step up in the hierarchy are selection processes that occur among cells or collections of cells within the body of an organism. John Buchholz (1922) has described a number of such processes that occur in vascular plants ranging from gametic and gametophytic selection through embryonic and interovular selection. He terms these intraorganismic processes *developmental selection*, and I will briefly describe just two types. In gametic selection competition occurs among gametes. For instance, among male gametes differences in the motility of sperms may affect the probability of fertilizing an egg. Such cases have been documented in plants (see Buchholz 1922; also see Lewontin 1970), and the potential is certainly there in animals (see Mulcahy 1975). Another interesting case is that of embryonic selection. Here there is a competition among embryos. Buchholz mentions several cases. For instance, among some conifers several embryos are formed and there is intense competition among them because only one will form the seed embryo. Compare this to the case where several seeds are formed in the same testa, as in citrus seeds. These seeds must germinate close to one another, resulting in severe competition. But here the competition is between seeds in the soil—in the external environment—and so, according to Buchholz, it is to be considered standard organismic selection. In other words, the line between developmental selection and organismic selection is not necessarily drawn in terms of differences in the level of organization of the interactor, but rather in terms of differences in the selective environment, or alternatively in terms of the stage of development of the interactor. In some cases this line may well be fuzzy. From our point of view, the important point is that gametes and embryos (as well as gametophytes and ovules) can be interactors within the organismic environment.

Leo Buss (1983) has described a selection process he terms *somatic selection*. This, like Buchholz's developmental selection, is a process that occurs within individuals, but unlike developmental selection this process results from the interaction of parts of individuals with the external environment. Within-individual variation would be of no evolutionary significance if Weismann's doctrine of the separation of the germ line from the soma were generally true. According to this doctrine, changes in the soma cannot be transmitted to the germ line. Thus, for instance, there could be no effective selection among different branches of a tree because no matter how they differed they would all produce the same seed genotypes. However, as Buss points out, this doctrine does not hold for any protists, fungi, or plants and is totally unsupported for nineteen phyla of animals. In organisms where there is no separation of the germ line from the soma, where all cells are capable of reproducing, somatic mutations can lead to within-individual variation on which selection can act. For instance, if a mutation occurs in a limb bud on a tree and is incorporated in the resultant branch, then differences between that branch and other branches of the tree may be of selective significance. If that mutation results, say, in a different color flower and this attracts more pollinators, then intraindividual interbranch selection will occur. Furthermore, the fortuitous mutation will be present in the seeds of the selected branch. (See Klekowski 1976, 1979, 1982, and 1984 and Mishler 1988 for dicussions of the evolutionary significance of somatic selection in ferns and mosses, respectively. Gill and Halverson 1984 discuss intraindividual interbranch selection in trees.) Again it should be clear that those parts of individuals between which somatic selection occurs are interactors.

Skipping the organismic level, we now turn to the possibility of selection at the level of groups. Wade (1977) has created group selection in a laboratory setting. Group selection in nature is more controversial (see Wilson 1983b for an illuminating discussion of this controversy; see Wilson 1983a for a plausible case of group selection in nature). So far I have not attempted to answer the empirical question of how prevalent group selection is in nature; rather the discussion was intended to shed light on the conceptual question of what should count as group selection. But for present purposes the important point is that when there is selection at the level of groups, these groups are interactors.

The groups pertinent to discussions of group selection are relatively small and short lived. Can selection occur at higher levels of organization, for example at the species level? There have been many recent discussions of species selection (e.g., Stanley 1975, 1979; Gould and Eldredge 1977; Eldredge and Cracraft 1980) but most do not distinguish clearly between mere differential replication of species and true species selection. That is, in the useful terminology of Vrba and Gould (1986), they do not distinguish between *sorting* and *selection*. Clearly there is sorting at the species level, that is, there is differential replication of species. But is there species selection? John Damuth (1985) argues that in general species are not the right sort of entity to participate in a selection process. In cases of organismic selection, selection occurs among organisms inhabiting a common selective environment. Note that the population consisting of these organisms, that is, the organisms inhabiting a common selective environment, is not necessarily a population united by gene flow, that is, a deme. According to most proponents of species selection (e.g., Stanley 1975), species selection occurs among species within a *clade*. But a clade is a genealogical rather than an ecological unit; and it is implausible that, in general, the constituents of a clade share a common selective environment.[13] Likewise, different local populations of a species

[13] If there is a hierarchy of interactors, then there is a corresponding hierarchy of

will often times not share a common selective environment. Thus species in clades are not analogous in the relevant way to organisms within a population of organisms inhabiting a common selective environment. In the first case we have a unit held together by gene flow (species) within a larger genealogically characterized unit (clade). In the second case we have an interactor (organism) within a population of interactors united by common selective forces. To have an explanatory hierarchical theory of selection we need a hierarchy of the right sort of units. Units of selection need not, and usually do not, correspond to units of a genealogical nexus. Damuth argues that local populations of a species within an ecological community could be units of higher level selection; and the community, not the clade, would be the unit within which selection occurs. Again, it should be clear that these higher level entities (what Damuth calls *avatars*) are interactors.

Nothing in Damuth's argument precludes species selection; his argument is that species *qua* units of gene exchange are not necessarily the sort of entity that participates in the selection process, that is, interactors. But in some cases there may be differences in species level properties that do lead to differences in speciation and extinction rates. For instance, in marine gastropods, species with planktotrophic larvae have greater larval dispersal compared to non-planktotrophs and so have greater gene flow. This process may lead to lower levels of speciation and extinction and therefore represent a genuine case of species selection (see Jablonski 1986 and references therein). Indeed Jablonski presents tantalizing evidence that during the end-Cretaceous mass extinction selection occurred at even higher levels of organization. Clades with more extensive geographic ranges showed higher survivorship than clades with smaller ones. Because individual species' ranges had no effect on clade survivorship, this can be considered an emergent property of clades.

selective environments. It is virtually inconceivable that all of the organisms in a clade would share a common organismic selective environment. But the relevant question here is whether or not species in a clade share a common species-level selective environment. Damuth's point is that there is no reason to expect that they do in general. This point holds even if it is not wildly implausible that in some instances species in a clade do share a common (species-level) selective environment.

I have argued that selection requires common selective environments. Selection at higher levels, for example, species and clades, requires much more extensive spatio-temporal selective environments. If mass extinctions are indeed caused by catastrophic events, such as the impact of large meteors, then this may have the effect of homogenizing vast parts of the earth with respect to certain selective pressures. If this is so one would expect increased higher-level selection during such periods.

I have offered a definition of levels of selection, and in this section we have discussed various levels at which selection may occur. It may occur among bare lengths of RNA within a "primordial soup," among lengths of DNA within cells, among chromosomes within cells, among parts of organisms within organisms (e.g., gametes in developmental selection, limbs in somatic selection), among organisms within (selectively homogeneous) populations, among groups of organisms within local populations, among local populations within communities, among species within groups of competing species (which may or may not correspond to clades), and finally among clades. I have argued in each case that when there is selection at a given level the entities at that level are interactors. This should not be surprising, since my definition of levels of selection is designed to pick out levels of interaction. Thus we have a hierarchy of interactors ranging from bare lengths of RNA through organisms to clades. Let me reemphasize that some of these interactors may also be replicators. The point is that when selection occurs at a given level the entities at that level must be interactors.

The hierarchy presented here apparently differs from that presented by other authors (e.g., Lewontin 1970; Hamilton 1975; Wimsatt 1980, 1981; Arnold and Fristrup 1982; and Wade 1984).[14] All these authors agree that in cases of organismic selection the

[14] I say "apparently" for the following reason. If you assume that every copy of a particular genotype has the same phenotype, then in standard cases of organismic selection there would be some genetic unit such that the variance in fitness at that level would be context independent. But that assumption is not likely to be true for any real population. When the assumption fails, copies of a given genotype could be partitioned by phenotype in a way that would be relevant to fitness. Thus the variance in fitness at the genotypic level would not be context independent. If all this is correct, I have no argument with Wimsatt's analysis; my argument would be against his application of that analysis.

"unit" of selection is some genetic unit—a gene, an entire chromosome, or even the entire genome depending on the amount of epistasis and gene linkage. For illustration I will present only Wimsatt's (1981, p. 144) definition of units of selection, but I believe that it applies to all of the aforementioned works:

> A *unit of selection* is any entity for which there is heritable *context-independent* variance in fitness among entities at that level which does not appear as heritable context-independent variance in fitness (and thus, for which the variance in fitness is *context-dependent*) at any lower level of organization.

Wimsatt, following Mayr (1963) and Lewontin (1974), argued against those (e.g., Williams 1966) who claimed that in standard cases of selection—those which we would classify as organismic selection—genes are the units of selection. Wimsatt's argument is based on certain general facts about genetic systems, most important among which is the fact that genes in general interact in the way they affect the phenotype. In particular, a given gene's effect on fitness depends on its genetic context. Thus the variance in fitness at the level of genes is, in general, not context-independent. Wimsatt concludes, again in agreement with Mayr and Lewontin, that the unit of selection in standard cases is a much larger genetic unit, that is, the entire genome. At other levels of selection, however, Wimsatt's definition agrees with mine. Thus when I say there is group level selection, Wimsatt would say that groups are the units of selection; when my approach implies that there is selection at the level of lengths of DNA, Wimsatt would say that these lengths of DNA are units of selection. Recall that such "selfish DNA" acts as an interactor. In fact, Wimsatt's analysis agrees with mine when and only when his units are interactors.

The hierarchy that results from approaches such as Wimsatt's lacks coherence. It includes replicators *qua* replicators and interactors *qua* interactors. Interactors and replicators play different roles in the process of evolution by natural selection. To resolve the "units of selection controversy" we need to ask coherent questions, such as: At what levels do the interactions between biological entity and environment that lead to differential replication occur? That is, what are the levels of interactors? My analysis is designed to answer this question through the use of a hierarchy of interactors ranging from bare lengths of RNA to entire clades. This hierarchy may not be exhaustive; there may be still other levels of selection we have not considered.

## 3. 4. A HIERARCHY OF REPLICATORS?

We might ask a second coherent question, one that has to do with levels of replicators: Is there an interesting hierarchy of replicators corresponding to the hierarchy of interactors? Let us examine the selection scenarios discussed above to determine what the replicators are in each case. In the case of the model of Eigen et al. (1981), bare lengths of RNA interacted within a "primordial soup." Differences in their physical structure resulted in differences in survivorship and rates of replication. Thus they are interactors. But it is obvious that they are replicators as well; indeed, they are paradigm cases of entities that replicate their structures directly and accurately. In the case of "selfish genes" (Doolittle and Sapienza 1980; Orgel and Crick 1980), lengths of germ-line DNA interact within a cellular environment so that some lengths dramatically increase their representation within the genome. Again, these lengths of DNA clearly are interactors as well as replicators, as is also true of lengths of chromosomes or whole chromosomes in cases of meiotic drive.

The cases falling under the rubrics of developmental selection (*sensu* Buchholz) and somatic selection (*sensu* Buss) are rather diverse. For instance, gametes may be interactors in some cases of developmental selection but they are not replicators. They do not replicate their structures directly but combine with other gametes to form diploid zygotes which, after a period of development, produce gametes. Thus the replicators are genes in this case. In somatic selection replicators are determined on the basis of sexual or asexual reproduction by the somatic parts. Here the replicators are the same as the replicators in ordinary organismic selection.

The next step in our hierarchy of interactors is the level of organismic selection. Here organisms are interactors, but what are the replicators? The answer again depends on the mode of reproduction. In sexual reproduction, the genomes of organisms are broken up by segregation and recombination; thus only parts of the genome reproduce their structure directly and accurately.

What do we call these parts of genomes? G. C. Williams defines a gene as "that which segregates and recombines with appreciable frequency" (1966, p. 24). (The strength of selection determines what counts as an appreciable frequency.) So the replicators in cases of organismic selection with sexual reproduction are genes (*sensu* Williams). In cases of asexual reproduction the entire genome is passed on directly from parent to offspring, and we could say that the entire genome is the replicator. Note that in asexual organisms the genome is a "gene" in Williams's sense. One could argue, as has Hull (1981), that in asexual reproduction the organism itself is a replicator. It replicates its structure in a fairly direct and accurate manner. In this case the difference between a whole organism and its genome is one of degree, and the vagueness of the notion of replicator allows us to say that either can be a replicator.[15]

The replicators in cases of group selection also depend on the nature of the reproductive process. At the risk of oversimplification we can distinguish two basic types of group selection processes, *intrademic* and *interdemic*. In cases of intrademic group selection, groups are formed during part of the organismic life cycle (e.g., the larval stage), and fitness-affecting interactions occur within these groups. Then the group members disperse into a common mating pool. The process of group selection occurs by differential group dispersal caused by differences in group structure (usually represented in formal models by different relative frequencies of alleles or genotypes). But group structure is not passed on directly to the next generation of groups; individuals from all the different groups unite in a common mating pool and sexually reproduce. New groups are formed in the next organismic generation at the appropriate stage in the life cycle. The replicators here are simply the replicators in normal sexual reproduction, namely genes (*sensu* Williams). In cases of interdemic group selection, groups are more or less reproductively isolated. Organismic selection occurs within groups and group selection occurs between

[15] I will not try to resolve this ambiguity, but to do so one would need to carefully distinguish various processes of asexual reproduction. For instance, in the asexual production of seed I would say that the genome is the replicator, whereas in reproduction by budding or fission the organism is the replicator. Some of these issues will be discussed in greater detail in section 3.6.

groups by processes of differential group extinction and propagation. Here the replicators are the groups themselves (which is not to say that genes are not also concurrently replicators with respect to the process of organismic reproduction). Group reproduction is a splitting process more similar to ameboid or bacterial reproduction than to sexual reproduction in higher plants and animals.[16]

Cases of what Damuth calls *avatar selection* (selection among local populations of species within an ecological community) and other, even higher-level, selection processes (species selection, clade selection?) are similar to interdemic group selection in that the group itself (avatar, species, clade) is the replicator, because the reproductive process is a splitting process of these higher-level entities. Thus we get the dual hierarchy of interactors and replicators listed in table 3.1.

I want to make three points about this dual hierarchy. The first

TABLE 3.1. HIERARCHIES OF INTERACTORS AND REPLICATORS.

| Selection Scenario | Interactor | Replicator |
|---|---|---|
| Origins of life | Lengths of RNA | Lengths of RNA |
| "Selfish genes" | Lengths of DNA | Lengths of DNA |
| Meiotic drive | Chromosome (or a part thereof) | Chromosome (or a part thereof) |
| Developmental or somatic selection | Parts of organisms | Genes or genome |
| Organismic selection: asexual reproduction | Organism | Genome (or organism?) |
| sexual reproduction | Organism | Genes |
| Intrademic group | Group | Genes |
| Interdemic group | Group | Group |
| Avatar selection | Avatar | Avatar |
| Species selection | Species | Species |
| Clade selection | Clade | Clade |

[16] Group reproduction and the differences between interdemic and intrademic group selection will be discussed in greater detail in sections 3.5–3.7. Also see the introduction to part 3 in Brandon and Burian (1984) or Wade (1978), reprinted therein.

is that the hierarchy in the interactor column is a fairly neat hierarchy of inclusion, whereas the replicator hierarchy is not so neat.[17] This hierarchy could be made to look neater if we adopt Williams's abstract notion of gene mentioned above. In that sense of "gene," all the replicators up to the case of interdemic group selection are genes. But that neatness is illusory if we think of the hierarchy as one of *physical* inclusion.

My second point is that the replicator hierarchy is derivative from the interactor hierarchy in the sense that we need to determine the level of interaction in order to determine the level of replication, but not vice versa. For instance, if we know that group selection is occurring, we can determine the appropriate replicators by the group reproductive process (intrademic versus interdemic selection). On the other hand, if we know that the replicators are genes, we do not know the level of interaction. (The relation between interactors and replicators is a one-many relation.) Because of this the first point is of little importance. Given a hierarchy of interactors we simply let the replicators fall where they may.[18] My final point is that single hierarchies (such as that presented by Wimsatt) that mix interactors and replicators serve to answer neither the question about interactors nor the one about replicators. Their lack of coherence clouds the real issues.

## 3.5. GROUP SELECTION VS. INDIVIDUAL SELECTION IN HETEROGENEOUS ENVIRONMENTS

Although the dual hierarchical approach to selection defended above is basically correct, in my opinion, it leaves a number of

[17] The interactor column is a hierarchy of inclusion if we ignore the first entry, that is, the case of selection among RNAs within a primordial soup. But since that process presumably occurred prior to, and not concurrent with, the others, there is a reason to ignore it when our concern is with a hierarchy of concurrent (or *possibly* concurrent) selection processes.

[18] One might compare this dual hierarchy with those presented by Eldredge and Salthe (1984), Eldredge (1985), and Salthe (1985). There is one major difference. The hierarchy I have presented is relative to a specific process, namely selection. Theirs is not. Thus Eldredge and Salthe take a broader view of interactors than I do. According to Hull's definition, which I have adopted, interactors imply selection. But there are many forms of interaction (mass-energy interchange) with the environment that do not necessarily lead to selection. Perhaps then my dual hierarchy is a special case of theirs, but it is just this special case that illuminates selection.

important questions unresolved. Some of these questions concern group selection and its relation to individual or organismic selection, and I will address them in this section and in sections 3.6–3.8. In section 3.9 I will briefly explore alternative approaches.

### 3.5a. Nunney on Individual Advantage

Historically—that is, in very recent history, say the last ten to twenty years—one of the primary catalysts to theoretical work on models of group selection has been evolutionary altruism. In evolutionary biology an altruistic behavior is defined as one that has a cost to the actor and a benefit to the recipient of the action, and costs and benefits are measured in terms of expected fitness. For example, Wynne-Edwards (1962) argued that many animals regulate their population size and density by reproducing at less than their capacity. This, supposedly, is a type of altruism, for organisms that reproduce at less than their capacity incur a cost in fitness and increase the relative fitness of those that do not decrease their reproductive rate. Other putative examples of altruism include predator alarm calls and the sacrificial behavior of worker ants. Whether or not any of these examples are genuine altruism is debatable (see G. C. Williams 1966), but if altruism does exist it presents a problem for evolutionary theory. Namely, how could such behaviors evolve? By definition, altruists are less fit than non-altruists, and so insofar as the behavioral traits are heritable, one would expect any mutant altruist always to be selectively eliminated. In response to this question many theoretists proposed models of group selection whereby altruism could evolve. (For reviews, see Wade 1978 and Uyenoyama and Feldman 1980.)

Consider the following simple model: There are two types in the population; type 1 does not affect the fitness of its neighbors, while type 2 does. Individuals randomly form small groups and fitness-affecting interactions occur within groups. Selection occurs within groups, then the survivors from all groups mix, and the cycle starts again. Thus this is a model of *intrademic* group selection or a *structured-deme model*. D. S. Wilson (1979) calls such groups *trait groups*. Type 2 individuals enhance the fitness of every member of their group, including themselves, by the amount $b$, but in so doing they incur cost $c$. In a group containing $x$ type 2 individuals the fitnesses are

$$W_1 = 1 + bx$$
$$W_2 = 1 + bx - c,$$

where $W_i$ is the fitness of type $i$. (This fitness function is from Charnov and Krebs 1975, but is quite standard in the literature.) Since $c > 0$, type 1 individuals are at an advantage within each group, but provided $b > c$ type 2's are at an overall advantage in the population as a whole. How can this be? Type 2's are altruists; their behavior decreases their fitness relative to type 1's in their group. But the group as a whole benefits from the presence of type 2's.[19] (To give this biological meaning, think of type 2's as organisms that give alarm calls when they perceive the presence of a predator. The alarm call increases the chance that the predator will notice the caller but decreases the overall chance that any member of the group will be caught by the predator.) Thus the average fitness of individuals within groups is greater in those groups with larger numbers of type 2's. For instance, suppose there are two groups, both of size 10, and that one group contains two type 2's and the other contains eight type 2's. Substituting 2 for $x$ we can calculate the average fitness of the individuals within the first group; it is $1 + 2b - 0.2c$. By a similar calculation we find that the average fitness within the second group is $1 + 8b - 0.8c$. Since $b > c$, $1 + 8b - 0.8c > 1 + 2b - 0.2c$. We can also calculate the overall fitness of either type within the population consisting of both groups. The overall fitness of type 1, $W_{o1}$, is $1 + 3.2b$, while $W_{o2} = 1 + 6.8b - c$. Clearly $W_{o2} > W_{o1}$, thus there is selection for type 2's.[20]

It is noteworthy that this process requires variation in the relative frequency of type 2's between groups. If there is no variation between groups, then the average fitness of all groups is the same and the overall fitness of type 1's exceeds that of type 2's. In effect,

[19] The behavior of type 2's decreases their *relative* fitness compared to type 1's. But since $b > c$, such behavior does not decrease their *absolute* fitness. This sort of altruism has been termed *weak altruism*, in contrast to *strong altruism*, where absolute fitness is decreased, for example, where $b < c$. See Wilson (1980).

[20] For present purposes I am not concerned with the evolutionary response to this form of selection, and so I have left the genetics out of this model. One could assume that type 1's and 2's were asexual haploids.

if there is no variation between groups, then the overall result of selection is identical to the result within any one group. Just as organismic selection requires variation among organismic phenotypes, group selection requires variation among group phenotypes.

When $b > c$ and there is sufficient variation between groups in the relative frequency of type 2's within groups, the frequency of type 2's in the population as a whole will increase.[21] Two conclusions are usually drawn from this sort of model: (1) The behavior of type 2's is individually disadvantageous, that is, it is altruistic; and (2) it spreads in the population as a whole by a process of group selection. Leonard Nunney (1985) rejects both conclusions.[22] He argues that the first conclusion is based on an improper comparison of the fitnesses of types 1 and 2. A proper comparison of types must take place in a common selective environment. This point, of course, is entirely congruent with the analysis of selection developed in chapter 2. Nunney asks us to suppose that trait groups are of size $n$. The comparison of $1 + bx$ with $1 + bx - c$ is, Nunney argues, a comparison of the fitness of type 1 in one environment with that of type 2 in a different environment. It is a comparison of type 1 in an environment of $x$ type 2 *neighbors* with type 2 in an environment of $x - 1$ type 2 neighbors. (The key here is that the target type 2 is not its own neighbor.) It is important to keep the composition of the $n - 1$ neighbors fixed and to compare the performance of types 1 and 2 within a common context. To make this clearer Nunney rewrites the fitness equations in terms of equivalent group neighbors:

$$W_1(x) = 1 + bx$$
$$W_2(x) = 1 + b - c + bx,$$

where $W_i(x)$ is the fitness of type $i$ with $x$ type 2 neighbors (Nunney 1985). If we accept these fitness functions, it follows that regardless of the number of type 2's in a group, type 2's are favored provided that $b > c$. Another way of thinking about this is to apply

[21] Wilson (1980) has shown that the random formation of groups produces sufficient between-group variation for the evolution of type 2 (weak altruism). Also see Uyenoyama and Feldman (1980) for the same result.

[22] John Maynard Smith would also reject these conclusions. See Maynard Smith (1976 or 1987). I will focus on Nunney's arguments since they are more powerful.

a "mutation test" (Matessi and Karlin 1984; Nunney 1985). Given the composition of the $n - 1$ group members, would the target individual increase its fitness by "mutation" from type 1 to type 2 or vice versa? When $b > c$, the individual is better off being a type 2, because a type 1 mutating into a type 2 nets a gain of $b - c$. Thus Nunney concludes that type 2 is not a form of altruism but a form of *benevolence*; it is a trait that is individually advantageous.

Although this argument sounds quite plausible, I believe that it is seriously flawed. First, the mutation test as presented above is not applicable to small populations. Nunney realizes this. In small populations the case of a type 1 with fitness $W_1(x)$ mutating to a type 2 with fitness $W_2(x)$ is handled as follows: "Type 1 has an individual advantage only if $W_1(x)$, relative to the average fitness of the population, is greater than $W_2(x)$, relative to the postmutation average fitness of the population" (Nunney 1985, p. 217). Let 0.1 and 0.09 be the values for the parameters $b$ and $c$, respectively, and suppose groups are of size 4, that is, $n = 4$. Since $b > c$, type 2 is fitter than type 1, according to Nunney's fitness function, and type 2 passes the mutation test in large populations (e.g., $N = 100$; this would be a population of twenty-five trait groups). But in a small population, say one of size 8 (two trait groups), type 2 fails the mutation test, that is, it is less fit than type 1. It follows that even when both the relative frequency of type 2's within groups and the size of groups are held constant, whether type 2 is individually advantageous or disadvantageous depends on the population size (the number of trait groups). Consider a population that consists of two trait groups of size 4, each containing two type 2's ($n = 4$, $x = 2$). According to Nunney's mutation test, here type 2 is not individually advantageous. Add ninety-eight identical trait groups to the population. Now type 2 passes the mutation test, so it is individually advantageous (see Appendix to this chapter). This is a strange result. By hypothesis, fitness-affecting interactions occur within trait groups and not between them. But this result means that one cannot determine whether or not type 2 is individually advantageous simply by looking at the target trait group, the group within which fitness-affecting interactions occur. Rather, one must look at the other groups within the population. In other words, whether or not type 2 is individually advantageous depends on the presence or absence of other groups *be-*

*tween which there are no fitness-affecting interactions.* Things get stranger still when one tries to apply the mutation test criterion of individual advantage to populations consisting of a single group (i.e., populations with no group structure). Here population size determines if type 1 or type 2 is individually advantageous. No matter which type passes the mutation test, type 1 will always be selected in such populations (Wilson 1980). Clearly this criterion of individual advantage is not applicable to populations with no group structure, although it seems such a criterion ought to be applicable here if anywhere.

A second and more fundamental problem with Nunney's argument is his conception of the environment. As should be evident from chapter 2, I am in complete agreement with the idea that proper fitness comparisons must be made in a common environment. This idea is not new; it is the basis of *common garden experiments.* The name comes from botanical experiments where different genotypes are planted in a common plot (either in the greenhouse or in the field) to compare fitnesses. Such experiments are by no means limited to plants; twentieth-century biology literature shows they have often been conducted on fruit flies and flour beetles. (Most common garden experiments on animals take place in laboratory environments. See Gill, Berven, and Mock 1983 for descriptions of such experiments on frogs conducted in the field.) Since both absolute and relative fitness can vary with environment, the only meaningful fitness comparisons are those done in common environments. As I argued in chapter 2, the notion of environment that interests us here is that of the selective environment, the arena within which selection occurs. It consists of all factors, both abiotic and biotic, that affect fitness. It cannot be identified with external environment (which dominates most work in ecology) because variation there is neither a necessary nor a sufficient condition for variation in selective environments (see section 2.5). In particular, in cases of frequency-dependent selection such as the one we are now considering, an organism is a part of its own selective environment. In our case a type 2 enhances the fitness of every member of its own group, *including itself,* by the amount $b$; in this way it is a part of its own selective environment. And so within-group comparisons between types 1 and 2 are comparisons within a common selective environment. This follows di-

103

rectly from the definition of homogeneity of selective environments defended in chapter 2. An area or population is homogeneous with respect to selection if different copies of the same type (type 1 or type 2) have the same expected fitness, which is true within a trait group. Thus such groups are selectively homogeneous. A copy of type 2 within a group with $x$ type 2's has a different expected fitness than a type 2 within a group with $x + 1$ type 2's. Thus populations consisting of members of trait groups with different relative frequencies of type 2's are selectively heterogeneous.

Another way of approaching this problem is to ask what constitutes the proper design of a common garden experiment when selection is frequency dependent. Nunney says that when frequency-dependent effects are experienced within a group of size $n$, one should compare the relative fitness of competing types by fixing the composition of $n - 1$ members of the group, then compare the performance of the competing types against that fixed background. This means that we create some number of replicates of the $n - 1$ members and place type 1 in one such replicate and type 2 in a different one. Note that when we deal with populations with either no group structure or no variation in group structure (i.e., all groups have exactly the same relative frequency of the competing types), this procedure introduces variation in group structure into the experiment when there is none in the natural system.[23] And of course, changes in between-group variation affect the evolutionary dynamics in structured demes (Wilson 1980). Although Nunney's mutation test asks which type is better for an individual, it is also correct, and more to the point, to say that the individual is faced with a choice between types of groups in which to live. That is, given that there are $x$ type 2 individuals within the trait group, the mutating individual is "choosing" between a group with $x$ and a group with $x + 1$ type 2 individuals. The latter groups are favored given that $b > c$. But note that the mutation test introduces between-group variation where there may have been none and invites comparisons between individuals that do not interact in any way. In contrast, if we treat the organism as part of its own selective environment, then we make within-group com-

[23] D. S. Wilson, personal communication.

parisons between individuals that do interact, and the procedure does not introduce between-group variation where there was none.

The problems with Nunney's approach may be made clearer by considering a hypothetical case, but one much closer to real life. Suppose we are interested in the relative fitness of a rare cannibalistic morph in salamander populations living in small isolated ponds. (Although cannibalism is certainly not altruistic, it is a plausible example of a trait that has frequency-dependent fitness effects.) Because they are rare, when the cannibalistic morphs do occur there is usually only one to a pond. Suppose pond populations are of size $n$. Nunney's approach would be to take two ponds, each with $n - 1$ noncannibals, and two experimental subjects, one a cannibal and the other a noncannibal. Place one subject in one pond, the other in the other pond, and then compare the performance of the two experimental subjects. (Of course, in a real experiment we would need replicates, but let us ignore that now.) Although the two subjects do not interact in any way, we could still compare their fitnesses. However, it is obvious that this comparison is biologically meaningless. In the absence of cannibals, our noncannibalistic subject may have a relatively high fitness, but that has no bearing on what its fitness would be in a pond with a cannibalistic morph. The meaningful comparison would be the within-pond comparison, the comparison between different morphs interacting in a common selective environment.

If we restrict fitness comparisons to those done within homogeneous selective environments, then we can easily construct a definition of individual advantage. *A trait is individually advantageous relative to some set of alternative traits and to some selective environment if and only if that trait's expected contribution to fitness is not exceeded by any of the alternative traits.* This criterion shows that weak altruism, type 2, is not individually advantageous in the situation modeled above.

### 3.5b. Environmental Heterogeneity and Group Structure

Rejecting Nunney's conclusion that weak altruism, or benevolence, is individually advantageous removes one argument that intrademic or structured-deme models do not model group selection. That argument states that since benevolence is individually advantageous, the process leading to its evolution is simply individual selection.[24] But there are still some reasons to doubt whether such a process is group selection, because of the very close analogy between this process and the process of selection in heterogeneous selective environments (what I termed "compound selection").

Consider the case described in chapter 2 (section 2.4) of different genotypes randomly distributed over a patchy environment. There are two genotypes, $G_1$ and $G_2$ and two selective environments, $E_1$ and $E_2$. $G_1$ is fitter than $G_2$ in both environments and both genotypes are fitter in $E_2$ (see figure 2.8). But *by chance* the two genotypes are not distributed equally into the two environments; $G_2$ is much more abundant in $E_2$, that is, it is more abundant in the favorable environment. As we saw, the single generation outcome to this process would be the increase of the relative frequency of $G_2$. (If the unequal distribution is maintained, this result will recur over a number of generations.) Within each environment there is selection for $G_1$, but there is a between-environment component to the change in genotype frequencies as well; $E_2$ contributes more individuals to the next generation. Since $G_2$ is disproportionately represented in $E_2$, this results in the increase in relative frequency of $G_2$. $G_2$, of course, is individually disadvantageous.

This sounds remarkably similar to the intrademic model discussed above. Types 1 and 2 are randomly distributed into trait groups. Trait groups with different relative frequencies of types are by our analysis different selective environments. Type 1 is favored in all selective environments. But groups with higher percentages of type 2's are favored over groups with lower percentages. Thus there is a between-group, a between-environment, component that causes change in the frequencies of the types in the total population. This process results in the increase in relative frequency of type 2, a type that is individually disadvantageous.

These two processes are very similar. Are both of them cases of

[24] That argument would not be conclusive. Just because the trait is individually advantageous does not imply that it did not evolve by group selection.

group selection? Is neither? Or are there significant differences between the cases that allow us to draw a line between them, classifying one as group selection, the other not? I believe the case of selection in a heterogeneous environment described above is not a case of group selection. Here the fitness values of the genotypes are constant within each environment, that is, the fitness functions are not frequency dependent. Think of the genotypes as seeds and of the environments as patches of different soil types. The fitness of a genotype is determined by the soil it is in regardless of its surrounding neighbors. Selection occurs within soil patches in a predictable way, that is, $G_1$ has a higher realized fitness than $G_2$ in both patches. But because of the disproportionate number of $G_2$'s in the richer patch, $E_2$, the $G_2$'s increase in relative frequency in the metapopulation consisting of both patches. As I argued in chapter 2, this process is completely understandable as a process of individual selection occurring within selective environments and as a distribution process of the genotypes over the environments. There is no group selection here. On the other hand, I think the process modeled by intrademic or structured-deme models is genuine group selection. What are the important differences?

There are two. The first is that the increase in relative frequency of the disadvantageous genotype, $G_2$, would not be predicted by theory. If the two genotypes are randomly distributed over the patches, theory would predict proportionate distribution and thus the evolutionary increase in frequency of $G_1$. The process I described is essentially a form of drift. Of course, theory can predict that in finite populations drift is not unlikely, but it cannot predict the *direction* of its effect. In contrast, the evolutionary increase in frequency of the individually disadvantageous type 2 in structured demes is a predictable result of the model. But to say that the first result is unpredictable is not to say that it is unimportant in evolution. Sultan (1987) has argued that unpredictable environmental fluctuations are so ubiquitous among plants that the expected evolutionary response is the evolution of high levels of phenotypic plasticity (also see Bradshaw 1965). Gill, Berven, and Mock (1983) have documented the effects of unpredictable environmental variation on the reproductive success of red-spotted newts. Since this sort of effect has rarely been looked for, we can hardly judge its

107

evolutionary importance. Nonetheless, this difference between the two processes is significant; in only one case, the case of evolution in structured demes, do we get predictable evolution of individually disadvantageous types.

The second difference is even more important. Both evolutionary scenarios are cases of individual selection in selectively heterogeneous environments.[25] But in the case of structured demes the relevant environmental variable is group structure. Group structure, unlike soil type, is under selection and responds to selection. Groups with higher frequencies of type 2's send a larger number of individuals to the common mating pool than do groups with lower frequencies. Type 2's are disproportionately represented in such groups not by chance, but by the very nature of selectively favored groups—they are, after all, the groups with higher numbers of type 2's. The evolutionary result of this process is an increase in the relative number of groups with a high percentage of type 2's, or in other words, the evolutionary result in an increase in frequency of favorable selective environments. This is possible because in this case the organisms form their own selective environments. Formally this difference is marked by the fact that in the one case fitness is frequency dependent, and in the other it is not. When the population is structured into groups with varying relative frequencies of the types, selection can act on these groups. This is a higher level of selection, and the structure that creates it evolves in response to it.

This second difference between the two scenarios—that in one case the environmental variation (in soil) is independent of the evolving population whereas in the other case it is a part of that evolving population—explains the first difference. When group structure is the relevant environmental variable, we do not need chance to get a disproportionate representation of individually disadvantageous types in selectively favored groups. This fundamental difference warrants our drawing a line between the two pro-

[25] That evolution in structured demes is a case of individual selection in a selectively heterogeneous environment is not incompatible with its also being group selection. This is similar to meiotic drive. From the organismic point of view, meiotic drive is a form of inheritance bias. From the genic (or chromosomal) point of view, it is a case of selection. Arnold and Fristrup (1982) make clear that both descriptions are perfectly correct and therefore, obviously, compatible.

cesses and considering the latter process, that modeled by the structured-deme models, as group selection.

Nunney (1985) supports another argument against the above conclusion. According to this argument, group selection requires the nonrandom association of types (see Maynard Smith 1976, 1982, and Hamilton 1975). It is a well-established fact that strong altruism (i.e., a trait that increases the fitness of others at the cost of a reduction in absolute fitness) can evolve only when there is a nonrandom association of types (see Hamilton 1964 and Wilson 1980). But this finding is of dubious relevance if we recognize weak altruism to be individually disadvantageous.

A better way of appreciating this argument might be to put it in terms of individual mating patterns. These are the terms by which the distinction between interdemic and intrademic group selection is usually drawn. The groups in interdemic group selection are (more or less) reproductively isolated, that is, individuals from different groups never or rarely mate. On the other hand, in the intrademic models individuals group together during part of their life cycle, interact, and then disperse to mate randomly in the global population. These two patterns form the end points of a continuum. Following Wade and Breden (1981), we can describe this continuum as follows: a proportion of the global population, $f$, mates within groups, and the remainder of the population, $1 - f$, mates at random within the global population. Pure interdemic selection occurs when $f = 1$, and pure intrademic selection occurs when $f = 0$. $f$ can range from 0 to 1, and so we have a continuum based on the pattern of individual mating with pure intrademic selection at one end and pure interdemic selection at the other. Wade and Breden (1981) show that higher values of $f$ increase both the range of conditions under which altruism will evolve and its rate of evolution. But, recalling that weak altruism can evolve when $f = 0$, it is hard to justify drawing a line between group and individual selection within this continuum.

## 3.6. GROUP FITNESS

In chapter 1 I argued for the widely accepted conclusion that evolution by natural selection requires heritable variation in expected fitness (adaptedness). Selection among groups requires that they differ in actualized fitness and that this difference is due to differences in their adaptedness to a common selective environment. For this process to result in evolutionary change, these differences must be heritable. But what is group fitness, actual or expected? And what is group heritability? I will address these questions in this and the next section.

Independently, Damuth and Heisler (1988) and Mayo and Gilinsky (1987) have argued that there are two fundamentally different types, or at least two fundamentally different conceptions, of group fitness.[26] Damuth and Heisler distinguish between *multilevel selection [1]* and *multilevel selection [2]*, and draw the distinction in terms of models. In multilevel selection [1] models the focus is on the effect of group membership on *individual* fitness; thus group fitness here is the mean of the individual fitnesses within the group. In multilevel selection [2] models the interest is on the differential extinction and/or proliferation of groups; here group fitness may or may not correspond to the mean of the individual fitnesses within the group. They point out that population-genetic models of group selection (this includes both intrademic and interdemic models in the sense distinguished above) are multilevel selection [1] models, while species-selection models are primarily concerned with multilevel selection [2]. Although the distinction is drawn in terms of models and the interests of modelers, they suggest that this difference corresponds to a difference in group selection processes. The Mayo and Gilinsky distinction between their Type I and Type II models is, for our purposes at least, identical to Damuth and Heisler's. But Mayo and Gilinsky explicitly relate their distinction to a difference in natural processes and argue that only the process modeled by Type II models (i.e., processes where differences in group fitness cannot be identified with differences in individual survival and reproduction) is a distinct group-level selection process.

I want to consider the question of whether we can distinguish among group selection processes based on the nature of group fitness and its relation to the fitnesses of contained individuals. For the sake of this discussion, let us treat groups as higher level analogues of individual organisms; in particular, let group (expected)

[26] Also see Arnold and Fristrup (1982).

fitness be the *group's expected reproductive success*. This sounds sensible enough, but how do we determine group reproductive success? In order to determine reproductive success at any level of biological organization we need to be able to (1) individuate the entities at that level; and (2) trace parent-offspring lineages. When concerned with organismic selection, (2) may be difficult in practice, but it is conceptually clear-cut provided (1) can be accomplished. (1) is fairly easy among most animals. We can easily determine where one fruit fly ends and another begins, where one elephant ends and another begins. But among many plants the problem can be acute. Most plants can produce physiologically independent copies of themselves through a process of vegetative growth. For instance, a clump of wisteria can and will send out runners from which leaves and roots will sprout. When the runner is severed there is a new physically separate and physiologically independent plant. This process is often called "vegetative reproduction." John Harper (1977) has argued that this term is a misnomer, that this is simply a process of growth, not reproduction. Note that what counts as the reproduction of a plant depends on how we individuate plants. Harper argues that from the point of view of population biology, the multiple clones of a given wisteria genotype (what he calls *ramets*) are simply physically connected or disconnected—it does not matter—parts of the same *genet*. A genet consists of all the ramets descended from an original seedling (produced by sexual or asexual reproduction—again it does not matter), and it is the unit of plant population biology.[27] Thus, having

[27] See Dawkins (1982, pp. 253ff.) for an insightful discussion of the evolutionary significance of Harper's distinction between growth and reproduction. Dawkins's insight is that in reproduction the germ line goes through a single-cell bottleneck, and thus the developmental process of the organism must start anew. In growth from a multicell meristem, development does not start from scratch. Consider mutations or genetic rearrangements that fundamentally affect (i.e., affect the very beginnings of) the developmental process. In reproduction they can be expressed, and potential new adaptations can be selected. But in vegetative growth they would never be expressed (or would only be expressed in that part of the new ramet containing cells derived from the mutated cell), and so could not be selected. Thus reproduction, not mere growth, is a necessary part of the process of adaptation. Dawkins cites Bonner (1974) as the inspiration for this insight. Note that our description of evolution by natural selection as a *three*-stage process—a process involving interaction, replication, and development—is in essential agreement with Dawkins's point.

decided how to count plants from the perspective of population biology, that is, having settled (1), we know what counts as reproduction and so we can, at least theoretically, trace parent-offspring lineages.

I will suggest that some of the problems faced in trying to accomplish (1) and (2) with groups are analogous to the problems faced by plant population biologists. But let us start by assuming that we can individuate groups, that we can identify and count groups. Is there a difference between group fitness and the mean of the fitnesses of the individuals within the group? Are there cases where group fitness can be identified with mean individual fitness? Are there cases where it cannot be so identified? To answer these questions let us consider two extreme cases: (1) one in which there is variation in group fitness with no attendant variation in individual fitness; and (2) another where there is variation between groups in mean individual fitness but no variation in group fitness.

From a population genetic point of view, the first case may seem an impossibility.[28] How could groups differ in fitness if there is no between-group variation in individual fitness? Suppose there are two groups, A and B. At time $t$ both groups are of size $N$. Suppose there is no variation in individual fitness, either between or within groups, and that both groups are expanding geometrically so that at time $t + 1$ there are $2N$ A's; at time $t + 2$, $4N$ A's; at time $t + 3$, $8N$ A's, and so on. Group A continues to expand in this geometrical progression; but while the individuals in group B reproduce at the same rate, group B splits in two when it reaches size $2N$. Thus at $t + 1$ there are two groups, $B_1$ and $B_2$; at $t + 2$ these two groups split in two so that there are four B-groups; at $t + 3$, there are eight B-groups, and so forth. At time $t$ the A-groups and B-groups are in a 1:1 ratio; by $t + 3$ the ratio is 1:8, and it will continue to decrease. But the ratio of A-type individuals to B-type individuals is 1:1 and remains unchanged. And so B-groups have a higher group fitness than A-groups even though individual fitnesses do not vary between them.[29]

---

[28] Those committed to the view that all evolutionary change can be represented in terms of changes in gene frequencies would have to deny this possibility, because the frequency of the different genes associated with different groups would remain unchanged in the metapopulation composed of all the groups.

[29] This case is isomorphic to Arnold and Fristrup's (1982) case of two grasshopper

Is this group selection? First suppose these groups are part of some metapopulation of a species. It seems that B-groups do out-reproduce A-groups, but this proposition presupposes that what we have only *tentatively* identified as groups are *in fact* groups (i.e., it presupposes a certain approach to [1]). Our "groups," we may assume, are spatiotemporally distinct collections of individuals. But does that make them groups? What is a group? Uyenoyama and Feldman (1980, p. 395) have addressed this question and have offered the following definition of groups:

> A group is the smallest collection of individuals within a pop-ulation defined such that genotypic fitness calculated within each group is not a (frequency-dependent) function of the composition of any other group.

This definition accurately portrays the notion of groups at work in population-genetic models of group selection (Damuth and Heis-ler's multilevel selection [1] models). According to it, our "groups" are not groups at all, so there is no differential reproduction of groups and thus no group selection. Although this response may be question-begging in that it takes the perspective on group se-lection that is being called into question by our example, I believe that it is essentially correct. The case as we have described it is a higher-level analog of the following: Suppose there are two plant seedlings, one of genotype $G_1$, the other $G_2$. Both seedlings grow laterally by sending out runners, from which new leaves and roots grow. After a certain period of time the total biomass of the two genets and the total number of leaf-root modules are the same. But the runners in $G_1$ remain intact while the runners in $G_2$ die and rot away. Thus $G_1$ is a single physically continuous genet while $G_2$ is a genet consisting of physically discontinuous parts. But, as should be clear from the discussion above, this is not natural se-lection at the organismic level because there has been no differen-tial reproduction between the two genets. $G_1$ is, of course, the an-alog of the A-group and $G_2$ the analog of the B-group. Just as the physical splitting of a genet is not organismic reproduction, the

---

species speciating at different rates with no attendant variation in individual fit-ness. The difference is that their groups are species, genealogically independent entities, while our groups are not. This difference is important and will be dis-cussed later in this section.

physical splitting of a spatiotemporal collection of individuals is not necessarily group reproduction.

Let's change our example. Suppose genet $G_1$ is as described above but that $G_2$ now grows laterally by single-celled runners, and suppose that the somatic mutation rate is fairly high. Because the $G_2$ runners are single-celled, all somatic mutations are incorporated "downstream." As a consequence, $G_2$ produces a population of genetically diverse leaf-root modules on which selection can act. This form of vegetative "growth" is a form of reproduction.[30] A higher-level analog of the same process occurs when the A and B "groups" are different species (this is the Arnold and Fristrup case; see note 29). The A-species has high levels of gene flow that homogenize its local populations. The B-species has a much more restricted gene flow, and so its local populations tend to speciate. (To relate this to a widely discussed case of putative species selection: A and B are different species of marine gastropods; B has nonplanktotrophic larvae while A has planktotrophic larvae and thus greater larval dispersal and gene flow. See discussion and references above.) Over time the A-species remains intact while the B-species speciates into many B-type species. Since different B-type species are, by definition, genealogically distinct entities, the original B-species has produced a "population" of genealogically separate and diverse entities. This is reproduction, and so the B-species has outreproduced the A-species. Thus one necessary condition of selection, namely, differential reproduction, has been satisfied. Of course, this necessary condition is not sufficient; one could still argue along the lines of Damuth (1985) that this was not species selection because the species were not interactors within a common selective environment.

To summarize the conclusions drawn from our first extreme case: When we are dealing with within-population selection processes, there can be no variation in group fitness that is not reflected as between-group variation in mean individual fitness.

[30] According to Harper (1977, p. 27) the distinction between growth and reproduction is that "reproduction involves the formation of a new individual from a single cell: this is usually (though not always, e.g., apomicts) a zygote. In this process a new individual is 'reproduced' by the information that is coded in the cell. Growth, in contrast, results from the development of organized meristems. Clones are formed by growth—not reproduction." Also see note 27 above.

Thus in such cases group fitness is just the mean of the fitnesses of individuals within the group. On the other hand, when we are dealing with cases of selection between genealogically isolated "groups" (e.g., species, avatars?), group fitness need not correspond with the mean of individual fitnesses within groups. Here it is possible to have variation in group fitness with no attendant variation in individual fitness.

Now let's turn to the second extreme case, one where there is variation between groups in mean individual fitness, but no variation in group fitness. In particular, let us assume that we have two groups, A and B, that remain intact—that do not split and form new groups—over the time period under consideration. Further suppose that our groups are groups in the sense of Uyenoyama and Feldman (1980), that is, the fitness of individuals within groups is a frequency-dependent function of the composition of the group (and independent of the composition of the other group). Because of differences in group composition, the average fitness in one group is higher than that of the other group. Is this group selection? Damuth and Heisler (1988) would consider this group selection of their multilevel [1] type, and Sober (1984) would also class this as a case of group selection (see discussion in section 3.8). If they are correct, group selection is fundamentally disanalogous to individual selection in that the differential reproduction of groups would not be a necessary condition of group selection. Put another way, our case is one of the growth of one group relative to another. It is well understood why the mere growth of one individual relative to another does not count as individual selection. At the individual level at least, natural selection requires differential reproduction. That is not to say that differential growth may not be selectively relevant. It may well be, but the important point is that the faster growing individual may be selectively favored or *disfavored*, depending on the ecological situation.[31] As already noted, from an evolutionary point of view there is an important difference between growth and reproduction.

The case just described is a case of individual selection in a se-

---

[31] It is easy to imagine selective environments where larger individuals are selectively favored. But the reverse is equally plausible. For instance, if energy is a limited resource and if there is a trade-off between energy used for growth and energy used for reproduction, then the faster-growing types may well have lower fitness.

lectively heterogeneous environment. Like the Wilsonian intra-demic group selection models discussed above, the relevant environmental variable here is a group structure. But here, unlike the situation in the other models, group structures are not under selection because they are not reproduced. Without differential group productivity, individually disadvantageous traits cannot evolve (see Michod 1982, pp. 44–46, or Boyd 1982). Although in this case group structure affects individual fitness, group structure itself is not under selection (and so of course does not respond to selection). This selection scenario cannot lead to the evolution of group adaptations and is not a case of group selection.

From our consideration of the first extreme possibility—that of variation in group fitness with no attendant variation in individual fitness—we can conclude that (1) in cases of intrademic group selection, group fitness is identical to the mean fitness of individuals within a group; and (2) in cases of selection between genealogically isolated groups, group fitness is not identical to (and may not even be correlated with) the mean fitness of individuals within the group. From the discussion of the second extreme—that of variation between groups in mean individual fitness (when individual fitness is a function of group composition) with no variation in group fitness—we can conclude that (3) group selection requires variation in group fitness, that is, it requires differential group productivity. It follows that differences in group fitness, regardless of its relation to the mean fitness of contained individuals, drives the process of group selection.[32]

## 3.7. GROUP HERITABILITY

At the organismic level, evolution by natural selection requires the heritable variation of (expected) fitness. We might think that

[32] Mayo and Gilinsky (1987) define *group-individual* fitness as the mean fitness of individuals contained within a group, and *group-group* fitness as a measure of group reproductive success. I have not adopted their terminology, but I can restate my conclusions using it. Conclusion (3) is that true group selection requires differences in group-group fitness. Conclusion (2) is that sometimes group-group fitness is not identical to group-individual fitness. Mayo and Gilinsky argue for conclusion (3) and would accept (2) if the qualifier "sometimes" were dropped. But conclusion (1) is that sometimes (i.e., in cases of intrademic group selection) group-group fitness is identical to group-individual fitness. They fail to recognize this possibility and so rule out cases that I have argued are genuine cases of group selection.

the same would be true at the group level, that is, that evolution by group selection requires the heritability of group fitness. Maynard Smith (1987) has indeed argued for this conclusion, but Sober (1987) has argued against it. Surprisingly, neither articulates a concept of group heritability. For instance, Sober (1987) supports his position by reference to Wilsonian intrademic group selection, where he believes there is no group heritability. To me this is far from certain, but it seems fruitless to debate the issue without an explicitly defined concept of group heritability.

What *is* group heritability? I propose that it is the higher-level analog of the quantitative genetics notion of individual heritability.[33] Nowadays the standard population-genetic definition of heritability is the genetic one: the heritability of a trait is the fraction of the total phenotypic variance in that trait that is due to additive genetic effects (see, e.g., Roughgarden 1979, pp. 154ff.). This definition, however, assumes certain mechanisms of inheritance and is not applicable to cases where other mechanisms are operative (i.e., where replicators are entities other than genes; see Brandon 1985b for a discussion of heritability in cases of cultural transmission). It is a special instance of the generalized conception of heritability that originally comes from Galton (1889). The Galtonian notion is the one operative in quantitative genetics. It applies only to quantitative traits, that is, traits whose character states can be assigned to some metric. (This presumably includes most traits of selective relevance.)

What is it for a quantitative trait, such as height, to be heritable in a population? At an intuitive level, if height is heritable in a population, then taller than average parents have taller than average offspring and shorter than average parents have shorter than average offspring. This intuition forms the basis for the quantitative genetics definition of heritability, $h^2$. Suppose height is the quantitative character of interest here. Let $X_i$ be the height of individual $i$ and $X$ be the mean value of $X$ in the population. We are concerned with the relation between offspring and parental deviations from the mean. Let $X_o$ denote the average offspring value (average height) from parents whose midparent value is $X_p$. The midparent height is one half the sum of the heights of the *two* par-

[33] In making this basic proposal I follow Slatkin (1981). However, the details of my analysis do not follow his.

ents. This assumes sexual reproduction. If reproduction is asexual, then $X_p$ is simply the parent's height. The offspring deviation from the mean is $(X_o - \bar{X})$, and the parental deviation is $(X_p - \bar{X})$. A plot of offspring deviation against parental deviation (a scatter diagram) would yield a regression line whose slope is the heritability, $h^2$. The formula is

$$(X_o - \bar{X}) = h^2(X_p - \bar{X})$$

(Roughgarden 1979, p. 136, formula 9.1). Adding $\bar{X}$ to both sides yields

$$X_o = h^2 X_p + (1 - h^2)\bar{X}$$

(Roughgarden 1979, p. 136, formula 9.2). This formula tells us that the offspring phenotype depends on the parental phenotype to degree $h^2$ and on the population mean to degree $(1 - h^2)$. The values of $h^2$ range between 0 and 1. If $h^2 = 1$ the average offspring phenotype is just the (mid-) parental phenotype; if $h^2 = 0$ the average offspring phenotype is just the population mean. Thus the higher the value of $h^2$ the more offspring resemble their parents in deviations from the mean.[34]

To determine the heritability among entities at any level of biological organization one needs to be able to (1) individuate the entities at that level; (2) trace parent-offspring lineages; and, finally, (3) describe the relevant phenotypic variation in a quantitative fashion. In the last section we discussed some aspects of (1) and (2) with respect to groups. Further aspects of (2) will be explored later, and (3) presents no special problems. For the models of the evolution of altruism we have been discussing, the relevant group trait is the number of altruists in the group. Obviously, this trait is already described in quantitative terms. Although many group traits may be of potential relevance to group selection, in the following discussion we will focus on the number of altruists per group. In so doing we will lose no generality since our discussion will apply to any quantitative group trait.

[34] See Roughgarden (1979, chap. 9) for a discussion of the important limitations of this purely phenotypic definition of $h^2$. One that bears mentioning here is that $h^2$ is an empirical relation observed between parents and offspring in particular populations and particular ecological settings. Change the setting and $h^2$ may well change.

To trace parent-offspring lineages among groups we must first determine how groups reproduce. Following Wade (1978) I suggest that there are two basic ways in which groups reproduce, and a range of intermediate cases in between. In the first case a random selection of individuals from a group forms a "propagule" that is the beginning of a new group. There is no mixing of individuals between groups. (This is the method of group propagation used in Wade's experimental studies; see Wade 1976, 1977.) Here reproduction is a budding or fission process and the group-level analog of asexual reproduction. Group structure (e.g., relative number of altruists per group) is passed on directly from parent to offspring group, and between-group variation is not diminished by this reproductive process (Wade 1978). This is the "propagule pool model" of group reproduction. It is clear that this method of group reproduction is conducive to high levels of group heritability. Groups composed mainly of altruists will produce offspring groups composed mainly of altruists. The same is true for groups composed mainly of nonaltruists. Thus parental deviations from the mean group phenotype are directly transmitted to offspring groups.

The second method of group reproduction is the "migrant pool model." This is the means of group reproduction in Wilson's intrademic group (or trait group) selection model. Recall that in this model groups form during a part of the life cycle, and fitness-affecting interactions occur within groups. This leads to selection within and between groups, after which the members of groups disperse into a common pool, the migrant pool. The next generation of groups is formed by a random selection of individuals from that common pool. One might say that in this model groups do not reproduce at all—they come together during part of the organismic life cycle and soon after dissolve, and they are ephemeral entities not lasting long enough to reproduce their structure. This, perhaps, is Sober's (1987) reason for saying there is no group heritability in the case of intrademic group selection—if there is no group reproduction, there can be no group heritability. But I suggest that this is the group-level analog of sexual reproduction. The ephemerality of trait groups is no different than the ephemerality of the diploid stage in organisms where the haploid phase dominates the life cycle, as with many algae and mosses. In mosses we have no difficulty in thinking about selection among sporophytes (the diploid phase); that is, we have no difficulty in thinking about sporophytes as interactors. This is so even though sporophytes are ephemeral entities that do not directly replicate their structure. The difference between sporophytes and the diploid phases of higher plants and animals is one of degree, and we can certainly calculate the heritability of sporophyte traits. In Wilson's model the organisms dispersed to the migrant pool are the analog of the haploid phase of the life cycle. If groups are of size $n$, then $n$ of these haploids come together to form the $n$-ploid sexual phase of the cycle, that is, the trait group. The major difference between this and organismic sexual reproduction is that $n$ can be greater than two, so a daughter group can have more than two parents. Wade (1978) points out that this is analogous to blending (as opposed to particulate) inheritance. Blending inheritance has the effect of halving the population phenotypic variance and so is inimical to evolution.

Yet even with this form of reproduction, group heritability need not be zero. To calculate group heritability we need to compare the offspring deviation from the mean, $(X_o - \bar{X})$, to the parental deviation, $(X_p - \bar{X})$. How do we trace the parentage of an offspring group? If the offspring group is of size $n$, then each of the $n$ individuals can be traced back to a group in the last group generation (figure 3.1). Then $X_p$ is simply the mean of the phenotypic values, the $X_i$'s, of those $n$ parents.[35] Similarly, for each parental group we can follow its members through the migrant pool on into the next group generation to determine its offspring, whose average phenotypic value is $X_o$ (figure 3.2). Consider parental groups that deviate maximally from the mean. Such groups are composed of either all altruists or all nonaltruists. An all-altruist group sends only altruists to the migrant pool; likewise, an all-nonaltruist group sends only nonaltruists to the migrant pool. Offspring groups result from a random selection of $n$ individuals from the migrant pool. Because the selection is random, the migrants from all-altru-

[35] I am assuming either asexual or no organismic reproduction in the migrant pool. If there is sexual reproduction in the migrant pool, then a group of size $n$ has, potentially, $2n$ parents. This would complicate, but not qualitatively change, my conclusions.

Group generation 1

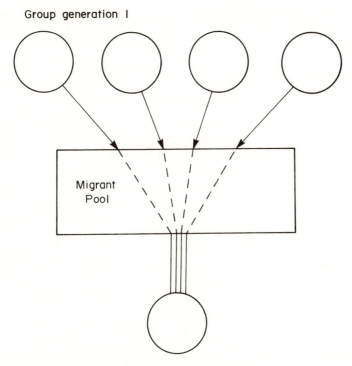

Group generation 2

Figure 3.1. Schematic diagram showing how to trace the parentage of an offspring group. Each individual in the offspring group can be traced back through the migrant pool to individuals in the parental groups.

ist groups will, on average, be put into the same range of samples of size $n - 1$ (the offspring group minus themselves), as will the migrants from all-nonaltruist groups. The migrants from the all-altruist groups will increase the number of altruists in each of the samples by one, while the migrants from all-nonaltruist groups will leave that number unchanged. In contrast, the mean offspring value would be the expected result of adding migrants from a group with the parental mean value to those samples of size $n - 1$. (For instance, if the relative frequency of altruists in the total population is 0.5, then half of these samples would increase their number of altruists by one, while that number would remain un-

121

Group generation I

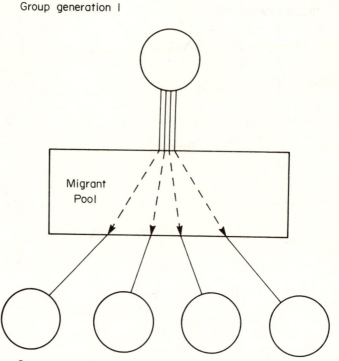

Group generation 2

Figure 3.2. Schematical diagram showing how to trace the offspring of a parental group. Each individual in the parental group can be followed through the migrant pool to individuals within offspring groups.

changed in the other half.) Thus offspring groups of all-altruist groups have more altruists than average, and offspring of all-non-altruist groups have fewer altruists than average. That means that group heritability is nonzero. Any method of forming groups from the migrant pool that would increase between-group variation would increase group heritability. One such method is incomplete mixing in the migrant pool. With less and less mixing, a migrant pool model merges into a propagule pool model.

Wilson (1980) has shown that the minimum conditions under which individually disadvantageous traits can evolve are those where trait groups are randomly formed. With random trait group formation weak altruism can evolve. As discussed above, a weakly altruistic trait is individually disadvantageous; it decreases the relative, but not the absolute, adaptedness of its possessors. Under such conditions group heritability is nonzero. Conditions that create greater between-group variation, such as those in the propagule model, also increase group heritability. Under these conditions strong altruism can evolve. In any case, the evolution of group adaptations requires group heritability.

## 3. 8. SOBER ON GROUP SELECTION

Much of the second half of Elliott Sober's recent book, *The Nature of Selection* (1984), is devoted to the development of a definition of group selection. His account differs significantly from the one developed here; some of these differences connect in interesting ways to some of the issues raised in the preceding three sections.

In one very basic respect Sober and I are in complete agreement: we agree that questions about the levels or units of selection are questions of causation. I have used the screening-off criterion to pick out levels of causal interactions. Sober uses what is presumably an independently motivated philosophical theory of causation to help elucidate some of the conceptual muddles that have arisen about the "units of selection" controversy.[36] Sober's analysis of causation is designed as an analysis of population-level causal claims—for example, that smoking causes lung cancer in the U.S. population; and not of individual-level causal claims—for example, that Smith's smoking caused his lung cancer. The analysis yields the following characterization of *positive causal factor*: C is a positive causal factor for E if and only if C raises the probability of E in at least one causally relevant background context and does not lower it in any (chapter 8). Thus smoking is a positive causal factor for lung cancer (in the U.S. population) if and only if smok-

---

[36] I have argued that there is no unitary question over the "units of selection"; there are questions concerning levels of selective interaction and questions concerning the replicators for any given level of selective interaction. Sober mentions the replicator-interactor distinction (1984, pp. 252–255) but does not really utilize it. However, for the most part I believe that when Sober talks about units of selections his primary concern is with the level or levels of causal interaction that lead to differential replication.

ing raises the probability of lung cancer for at least some smokers and does not lower it for any. Although this analysis of population-level causation may have its problems (see Gifford 1986), my major differences with Sober are not due to the analysis but to its application.

Sober (1984, p. 280) defines group selection as follows: "There is group selection for groups that have some property $P$ if (and only if)

1. Groups vary with respect to whether they have $P$, and
2. There is some common causal influence on those groups that makes it the case that
3. Being in a group that has $P$ is a positive causal factor in the survival and reproduction of organisms."

This definition attempts to provide necessary and sufficient conditions for the occurrence of group selection. I will argue that it provides neither.

The conditions are not sufficient because they do not require the differential reproduction and/or extinction of groups for group selection. Recall the case discussed in section 3.6 of one group *growing* relative to another because of between-group differences in individual mean fitness. In that case differences in group structure led to the differences in mean individual fitness. Let us label the group structure of the faster growing group $P$, a positive causal factor for survival and reproduction for the organisms in that group. Thus the case fits Sober's definition, which means that the mere growth of one group relative to another counts as group selection (when the differential growth is due to differences in group structure).[37] But as I argued in section 3.6, if this is group selection,

[37] Sober is aware of this consequence of his definition. After describing a possible experimental design in which the experimenter occasionally kills some percentage of the members of the smaller of two groups, Sober (1984, p. 318) writes: "This would be group selection in our sense—membership in a large group would be a positive causal factor in an organism's survival and reproduction—though it might turn out that no group founds a new colony and no group goes extinct." Later (p. 330) Sober defends this aspect of his definition: "Although the founding of new groups and the extinction of old ones is a prerequisite for altruism to evolve by group selection, it is not required by the definition of group selection given earlier. There are two reasons for this. The first is that I aimed at defining *group selection*, not *evolution by group selection*. The second is that I defined *group selection*, not *group selection for altruism*." I agree with both of these strategic remarks—indeed, I have

then this sort of group selection is fundamentally disanalogous to individual selection in that (1) it does not require differential reproduction; and (2) it cannot lead to group adaptations, whereas organismic selection can, of course, lead to organismic adaptations. These two points are really two sides of the same coin; the process of adaptation (i.e., the process of evolution by natural selection) requires differential reproduction. I am not arguing that one can deduce from some widely accepted conception of group selection that Sober's case is not group selection. There is no widely accepted concept of group selection. Rather, I am arguing that in defining group selection we ought to do so in a way that preserves the basic features of selection as we understand it in the organismic case. Differential reproduction is a fundamental feature of the selection process; in both the organismic and the group cases it is a necessary condition for adaptation at that level.

Nor does Sober provide necessary conditions for group selection. It is a consequence of his definition that for group selection to occur, every member of a group must be similarly affected (see pp. 319, 340, and 345). But why should this be the case? Individual selection can certainly occur at the expense of some of the selectee's parts. (Consider how a male praying mantis manages to pass his genes on to the next generation: he loses his head.) Why couldn't group selection occur at the expense of the fitness of some of the members of the group? Sober might look to the standard mathematical models of group selection for confirmation of his view. Recall the Charnov and Krebs fitness function discussed above:

$$W_1 = 1 + bx$$
$$W_2 = 1 + bx - c,$$

where $W_i$ is the fitness of type $i$, and $x$ is the number of type 2's in the group. Type 2's enhance the fitness of every member of their group, including themselves, by the amount $b$; but in so doing they incur cost $c$. One group is selectively favored over another if it has a greater number of type 2's (provided $b > c$). It is easy to see that in this selection scenario every member of a selectively fa-

---

followed them. My argument is that the case of differential growth between two groups is not *group selection*.

vored group benefits identically by the group selection component (the amount $bx$). This is more than is actually required by Sober's definition, which requires only that being in a selectively favored group be a positive causal factor in the individual's fitness. Some organisms could benefit more than others in a selectively favored group, and clause 3 of Sober's definition would still be satisfied. What is ruled out is that some organisms benefit and others are harmed by being in a selectively favored group. But this is precisely what can happen with only a minor modification of the above fitness function.

Suppose that instead of the situation modeled by the Charnov and Krebs fitness function, we again have two types: type 1 individuals do not affect the fitness of the other members of their group, while type 2's do. But in this case type 2's affect the other members of their group in a discriminatory way. They enhance the fitness of all type 2's in their group, including themselves, by the amount $b$, and in doing so they incur a cost $c$ per every individual aided. They behave in a way we might characterize as "spiteful" toward type 1's; that is, they decrease the fitness of type 1's in their group by the amount $s$, and in so doing they incur a cost $p$ per individual harmed. In a group of size $N$ containing $x$ type 2 individuals, the fitnesses are

$$W_1 = 1 - sx$$
$$W_2 = 1 + bx - cx - p(N - x).$$

In this case groups with higher proportions of type 2's are at an advantage under certain parametrizations.[38] But note that the advantage of being in a group with a high percentage of type 2's does not redound to type 1's. Indeed the higher the number of type 2's, the lower the fitness of type 1's. Thus type 1's are negatively affected by being in a selectively favored group.

I make no claims for the biological realism of the above fitness

[38] This minor modification makes the model more difficult to analyze. For groups with higher numbers of type 2's to be selectively favored, $b$ must be greater than $c$. Furthermore, when type 2's are rare, $b$ must be much greater than $c$, $p$, and $s$. For example, if $N = 10$ and $x = 1$, then for that group to be selectively favored over a group with no type 2's it must be the case that $(b - c) > 9(p + s)$. When type 2's are more common, the conditions for type 2's further evolution become less stringent. For example, if $N = 10$ and $x = 9$, then for the group to be favored over a group with no type 2's, it is necessary that $(b - c) > (p + s)/9$.

function, though it seems no less plausible than the Charnov and Krebs function. My point is that from a theoretical point of view it is possible for group selection to affect some group members positively and others negatively; furthermore, I see no biological reason why this theoretical possibility could not be realized. Therefore, the condition that all group members be similarly affected by group composition is not a necessary condition for group selection. Sober comes to this conclusion on the basis of clause 3 in his definition of group selection and by his definition of a positive causal factor.

It follows that Sober's definition gives neither necessary nor sufficient conditions for the occurrence of group selection. Where has he gone wrong? We could try to pinpoint the problem in his analysis of causation, but I think the basic problem lies elsewhere, in his use of organisms as the "benchmarks" of the group selection process.[39] That is, he sees group selection occurring if and only if group structure affects individual fitness in certain ways. (This is also the approach of Damuth and Heisler 1988 to the type of group selection they call multilevel selection [1].) But, as we saw in section 3.6, group fitness sometimes can and sometimes cannot be identified with mean individual fitness; in either case, differences in group fitness are required for group selection. Adopting this perspective, neither of the two problematic cases described above arise. The first case—the one of differential growth—would not be considered group selection because there is no differential *group* fitness. The second case—the one of the discriminatory altruists—could be seen as group selection because the group property P needs to be a positive causal factor for *group* survival and reproduction. This is true even though it is at the expense of some of the parts of the selected groups.

## 3.9. ALTERNATIVE APPROACHES

I have argued that the process of evolution by natural selection—the process of adaptation—requires both interaction and

[39] Sober also defines *genic selection* in terms of its effects on organisms, thus ruling out most of the real cases of genic selection, for example, Doolittle and Sapienza's "selfish genes." Later he allows that one might use benchmarks other than organisms for selection occurring at other levels of organization (pp. 308ff. for genic selection, p. 356 for groups).

replication. Sometimes one and the same entity plays both roles (e.g., "selfish genes"), but often that is not the case. Thus in standard cases of organismic selection, organisms are interactors and genes are replicators. I have presented a hierarchy of (potential) interactors. If we are concerned with the levels of selection, then we are concerned with this hierarchy. Corresponding to each level of interaction is a replicator. These replicators do not form a neat hierarchy, and the "hierarchy" of replicators that does result is derivative of the interactor hierarchy in that we can infer the replicators from the level of interaction (plus mode of reproduction) but not vice versa. Of what value is this hierarchical approach, and what are its alternatives?

The theory of evolution by natural selection is the only theory that can explain the origins and maintenance of adaptations.[40] If these explanations are to be scientific rather than mere exercises in storytelling, then adaptations must be carefully related to the selection processes that produce them (see Brandon 1981a and 1985a; see further discussion in chapter 5). For instance, early to mid-twentieth-century ecology and ethology are notorious for their explanations of organic features in terms of the features being "for the good of the species." Today we understand that species benefit is irrelevant to a selection process occurring at the level of organisms. But if there is a hierarchy of interactors, if selection occurs at different levels, then we cannot axiomatically assume that all adaptations are for the good of organisms, that is, we cannot assume that all adaptations are to be explained in terms of their organismic benefits. Indeed, it is just this assumption that Doolittle and Sapienza (1980) criticize as the "phenotype paradigm" (by "phenotype" they mean organismic phenotype). As they point out, it is futile to search for the organismic benefit of the repetitive sequences of DNA that they call "selfish genes." It is futile not because the organismic benefit does not exist (in this case it doesn't), but rather because of the irrelevance of any such benefit to the intracellular selection processes that produce these repetitive sequences. Similarly, the organismic benefit of any product of a

---

[40] There is an alternative explanation for the *maintenance* of apparent adaptations (i.e., features that increase the adaptedness of the relevant biological entities), namely, the competitive sorting of different phenotypes among different environments. See Horn (1979).

higher-level selection process, for example, group selection, is not directly relevant to the explanation of its origin and/or maintenance.[41]

I have not claimed that each level in the hierarchy of interactors is known to be of *evolutionary importance*. The importance of, say, somatic selection or group selection has not been conclusively demonstrated. Lacking such demonstrations, couldn't we argue that we should ignore consideration of hierarchical levels in applying the theory of natural selection to the biological world? I can think of only two such arguments, and both are seriously flawed. One is based on the assumption that we can know a priori that the only important level of selection is the organismic. In light of the conceptual and empirical work done during the last ten years, that argument cannot be taken seriously (see Brandon and Burian 1984 and Wilson 1983b). The other is based on bad methodology. The argument is that considerations of parsimony lead us to try to explain all adaptations as products of organismic selection (as suggested in Williams 1966). But once we see that other levels of selection are theoretically possible, we should not adopt a methodology that blinds us to their existence. It may well be that the only important, or the most important, level of selection is the organismic. But for the moment at least we need a hierarchical theory of interactors, if only to test the claim that organisms are the only important interactors in evolution.

If one recognizes the possibility of a hierarchy of levels of selection, are there any alternatives to the approach I have adopted? One might find many alternatives to some of the detailed positions of my analysis. For instance, one could certainly agree with my basic approach and still disagree with some of the substantive

[41] The last two statements of this paragraph are true but potentially misleading. Selection at one level can affect selection at other levels. For instance, genic selection for some repetitive DNA sequence could have negative effects at the organismal level, say be damaging or replacing important structural genes. Then negative organismal selection would alter the evolutionary trajectory of the "selfish gene." My points are conceptual and are meant to clarify the causal relevance of effects at one level of biological organization to selection at any level. The fundamental point can be put this way: when only one level of selection is in operation, only the effects at that level are causally relevant to that selection process; however, when multiple levels of selection are in operation, effects at one level can indirectly affect either higher or lower levels of selection. See Arnold and Fristrup (1982).

points I have made concerning group selection and its relation to individual selection. Or one could favor an alternative to screening-off as a means of distinguishing levels of selection. These would be variations on the theme I have presented. One could try to deny the relevance or utility of the interactor-replicator distinction in its application to questions concerning levels of selection. But because the failure to make this distinction has resulted in considerable confusion, I think such a move would be a step in the wrong direction. Finally, one could admit the relevance of the interactor-replicator distinction and the possibility of various levels of interactors, but argue that all of evolution by natural selection is understandable in terms of causes and effects acting at the genic level. This, I believe, is the position taken by Richard Dawkins (1982), which I find curious and, as I shall argue in chapter 5, ultimately untenable.

The model of intrademic group selection under consideration is as follows: Individuals form small groups and fitness-affecting interactions occur within groups. Selection occurs within groups, then the survivors from all groups mix and the cycle starts again. There are two types within the groups: type 1's do not affect the fitness of their neighbors, while type 2's enhance the fitness of every member of their group, including themselves, by the amount $b$, but in so doing they incur cost $c$. In a group containing $x$ type 2 individuals, the fitnesses are

$$W_1 = 1 + bx \tag{3.1}$$
$$W_2 = 1 + bx - c,$$

(where $W_i$ is the fitness of type $i$; Charnov and Krebs 1975). Nunney (1985) rewrites these fitnesses in terms of equivalent group neighbors:

$$W_1(x) = 1 + bx \tag{3.2}$$
$$W_2(x) = 1 + b - c + bx.$$

According to Nunney one can determine whether or not a trait is individually advantageous by a "mutation test." In a large population, type $i$ is individually advantageous if $W_i(x) > W_j(x)$ for all type $j$'s. In other words, keeping $x$ fixed, if an individual of type $j$ would be better off "mutating" to a type $i$, then type $i$ is individually advantageous.

This mutations test does not work in small populations because in such populations a single mutation can significantly affect the population mean fitness. In small populations the test is more complicated. Here we pick some target individual. Suppose it is a type 1. Then type 1 is individually advantageous only if $W_1(x)$, relative to the average fitness of the population, is greater than $W_2(x)$, relative to the *postmutation* average fitness of the population.

Consider the following set of parameter values: $b = 0.1$; $c = 0.09$; $n$ (group size) $= 4$; $N$ (population size) $= 8$; and $x = 2$. The Nunney fitnesses, that is, the fitnesses according to fitness function (2), are as follows:

$$W_1(x) = 1 + 0.1 \times 2 = 1.2$$
$$W_2(x) = 1 + (0.1 - 0.09) + 0.1 \times 2 = 1.21.$$

Assume there are two groups of size 4, both with two type 2's. Since both groups are the same, the population mean fitness (premutation $\overline{W}$) is simply the average fitness within one group, which is

$$\text{Premutation } \overline{W} = 2(1.2) + 2(1.2 - 0.09)/4 = 1.155.$$

Assume one type 1 mutates to a type 2. Then the postmutation $\overline{W}$ is

$$\text{Postmutation } \overline{W} = [4(1.155) + 1(1.3) + 3(1.21)]/8 = 1.1938.$$

(1.3 is the fitness of a type 1 in a group with three type 2's and 1.21 is the fitness of a type 2 in a group with three type 2's.) Now we can calculate $W_1(x)$ relative to the premutation $\overline{W}$ and $W_2(x)$ relative to the postmutation $\overline{W}$. They are

$W_1(x)$, relative to the premut. $\overline{W} = 1.039$, and
$W_2(x)$, relative to the postmut. $\overline{W} = 1.0136.$

It follows for this set of parameter values, and in particular when the population size $N = 8$, that type 1 is individually advantageous.

Now let $N = 100$ but keep the other parameter values the same. This means that we have twenty-five groups of size 4. Assume as before that all groups have two type 2's. The Nunney fitnesses remain unchanged:

$$W_1(x) = 1.2$$
$$W_2(x) = 1.21.$$

Again consider one type 1 mutating to a type 2. The premutation $\overline{W}$ is only the $\overline{W}$ in any group (since all groups are identical). Thus it is unchanged from the first case:

$$\text{Premutation } \overline{W} = 1.155.$$

To get the postmutation $\overline{W}$ we take the weighted average of the fitnesses of the four individuals in the group affected by the mutation and of the ninety-six individuals in the twenty-four unaffected groups. Thus,

Postmutation $\overline{W} = [96(1.155)$
$+ 1(1.3) + 3(1.21)]/100 = 1.1581$.

Again we can caculate $W_1(x)$ relative to the premutation $\overline{W}$ and $W_2(x)$ relative to the postmutation $\overline{W}$:

$W_1(x)$, relative to the premut. $\overline{W} = 1.039$, and
$W_2(x)$, relative to the postmut. $\overline{W} = 1.0448$.

Thus, when $N = 100$, type 2 is individually advantageous.

# The Structure of the Theory of Natural Selection

What is the structure of evolutionary theory? This question has received considerable attention from philosophers of biology. For instance, David Hull's (1974) textbook in philosophy of biology has a chapter called "The Structure of Evolutionary Theory." Eleven years later Alexander Rosenberg's (1985) book has a chapter with the same title. Elisabeth Lloyd's (1988) recent book is entitled *The Structure and Confirmation of Evolutionary Theory*. In 1980 the Philosophy of Science Association devoted a symposium to the topic (Beatty 1981; Brandon 1981; Williams 1981). And, of course, there have been numerous papers written on this issue. Thus it is safe to say that the topic is not unexplored. But evolutionary theory, broadly construed, includes theories of speciation, biogeography, evolutionary ecology, ecological genetics, population genetics, and phylogenetic systematics. It draws on developmental biology and biomechanics (though insufficiently, in the view of some) as well as paleontology and molecular biology. Perhaps it is not surprising that philosophers have not dealt with the interrelations of all these theories or areas when they have tried to describe "the structure of evolutionary theory." Indeed, in the philosophers' sense of the term "theory," I do not think evolutionary theory is a theory at all. Rather, it is a family of theories (and goals, methods, and metaphysics) related in complex and ever-changing ways. The theory of natural selection is only a part of this complex, but it is the part on which philosophers have typically concentrated their attention.

At least part of the reason for that concentration is the notorious "tautology problem"—the question of whether or not the principle of natural selection, or the principle of the "survival of the fittest," is a tautology. In this chapter we will devote some time to that question. Not only do I hope to convince the reader of the correctness of my proffered solution, but also—and this will be more difficult—I hope to show that the problem and its solution are genuinely interesting. But this solution is only a way station en route to a general structural description of the theory of natural selection. This general structural description will, I believe, deepen our understanding of the theory of evolution by natural selection, the theory that provides the only available explanation of the origins and maintenance of adaptations in nature.

## 4.1. EMPIRICAL CONTENT OF THE PRINCIPLE OF NATURAL SELECTION (PNS)

Is the PNS a tautology or does it have empirical content? In some circles (e.g., among most philosophers of biology) one would have to apologize for raising this question. "That's not a problem" or "That problem was solved years ago" are the vehement declarations one is likely to hear. But I refuse to believe that a problem has been solved when so-called experts give different and incongruous, if not inconsistent, answers. Furthermore, I believe that my own answer will be of interest to both general philosophy of science and evolutionary biology. But before jumping into it, let me begin by saying what the problem is *not*.

Creationists and the occasional Harvard lawyer (Macbeth 1971) like to argue that the alleged tautologousness of the PNS invalidates all of evolutionary biology. The suggestion is that a logician can by the casual analysis of a statement or phrase topple the entire edifice of Darwinian theory. Although many people have taken this threat seriously (see Lovtrup 1975 or Peters 1976, 1978), the suggestion is plainly silly. If the "tautology problem" is whether or not an entire scientific discipline can be destroyed by the logical analysis of the phrase "survival of the fittest," then indeed the "tautology problem" is a pseudoproblem. But that is not the problem to be dealt with here. I assume that Darwinian (or neo-Darwinian) evolutionary biology will survive our analysis of the PNS. But that is not to say that our understanding of the theory will not be affected by the analysis.

If the question is whether or not the PNS is a tautology, then there are two prerequisites for its resolution. The first is a clear statement of the PNS, that is, a clear statement of just what is or is not supposed to be tautologous. Not everyone who has ad-

dressed this question has bothered to do this. Creationists, jour-
nalists (Bethell 1976), the aforementioned Harvard lawyer, and
philosophers naive about biology (e.g., Smart 1963 and Popper
1974) have been content to analyze the phrase "survival of the fittest." That phrase, of course, not being a declarative sentence is
not even a candidate for being a tautology. I have devoted much
of chapter 1 to giving a precise statement of the PNS so I have
completed this first prerequisite.[1] The second prerequisite is to develop a clear notion of tautologousness, or in contrast, empirical
content. As Elliott Sober has pointed out, this is not entirely
straightforward.[2]

The word "tautology" has narrower and wider meanings. In
logic a tautology is a theorem of the propositional calculus (see
Mates 1972), that is, a tautology is a sentence that is true in virtue
of its truth-functional structure. The truth-functional structure of a
sentence is determined by the structure of its truth-functional con-
nectives (e.g., "not," "and," "or," "if . . . then"). For example,
"Creationists are silly or creationists are not silly" and "Sex is advantageous in heterogeneous environments or sex is not advantageous in heterogeneous environments" are both of the form "P or
not-P" and are true in virtue of that logical form. Obviously, empirical testing of these statements is both impossible and irrelevant
to a determination of their truth or falsity. In its broader meaning
a tautology is any statement that is true solely in virtue of the
meaning of its component terms (whether they be logical terms or
not), for example, "All hermaphroditic plants have both male and
female reproductive parts." Here again empirical testing is impossible and irrelevant.

It is quite natural to think that one can draw a hard and fast
distinction between statements that are true or false solely by virtue of the meaning of their component terms and those that are
true or false by virtue, ultimately, of the way the world is. This is
known as the *analytic-synthetic distinction*. As we have seen, for the

---

[1] For similar statements of the principle of natural selection, or the principle of
the survival of the fittest, see Waters (1986). My version comes from Brandon
(1978).

[2] Sober (1984, chap. 2) does a good job of setting out the relevant philosophical
concepts in much greater detail than I shall. For those who want this detail I highly
recommend Sober's chapter.

former statements empirical testing is both impossible and irrelevant. Intuitively, for the latter empirical testing is possible and is the only way of determining truth or falsity. This distinction is one of the cornerstones of logical positivism. The positivists used this distinction both in their theory of explanation and in their criteria to demarcate science from pseudoscience.[3] Unfortunately, they were never able to develop theories of empirical meaningfulness that satisfied their own criteria of adequacy (see Scheffler 1963, chap. 2, or Hempel 1965, chap. 2). Moreover, Quine (1951, 1960) has argued that the analytic-synthetic distinction could not be coherently maintained. For many this argument dealt a deathblow to logical positivism. I will not rehearse Quine's arguments, but the following may make their general thrust clear. According to Quine, empirical evidence may impinge on any statement in our conceptual system so that no statement is immune to revision. Thus, for instance, we may one day discover weird plants that lead us to revise the concept of hermaphroditism so that the statement "All hermaphroditic plants have both male and female reproductive parts" will no longer be accepted as analytically true.[4] Imaginative botanists may be able to come up with such a scenario. (Although, we must admit, it is much more difficult to imagine the empirical circumstances that would lead us to reject "2 + 2 = 4.") Many of those who reject the analytic-synthetic dis-

[3] The fundamental distinction for the positivists was the one between cognitive significance and nonsense. Statements were cognitively significant if they were either analytically true or false or were contingent and connectable to experience in such a way that their truth or falsity could be empirically determined. The residue was nonsense. Thus statements that were neither analytic nor testable in any fashion were nonsense and not a part of empirical science.

[4] The following imaginary history of botanical concepts serves to illustrate this point. Suppose we knew of only two sorts of flowering plants: dioecious or unisexual plants, that is, plants that are either male or female but not both; and plants that contain both male and female reproductive parts in the same flower. The latter we define as hermaphroditic. Thus it is true by definition (analytically true) that all hermaphroditic plants have both male and female reproductive parts in the same flower. Later we discover nondioecious plants whose flowers are not hermaphroditic; that is, plants whose flowers are all unisexual (either male or female), but where individual plants contain both male and female flowers. These plants do not fit into our conceptual system, so we expand the concept of hermaphroditism to include not only plants whose flowers are hermaphroditic, but these newly discovered plants as well. Under our new conceptual system, what once was analytically true is now empirically false.

tinction follow Quine in adopting a form of holism with respect to the testing of scientific theories. According to this view, individual components of a theory are not tested piecemeal; rather, the theory as a whole is subjected to empirical tests, and negative results can be dealt with in a variety of ways. (Holism disallows the possibility of *crucial* experiments.)

All this notwithstanding, the distinction between testable hypotheses and statements for which empirical testing would be both impossible and irrelevant is vital for understanding the *practice* of science.[5] Scientists do not treat theories as undifferentiated wholes, nor should they. They do, and should, design experiments to test specific hypotheses.[6] I suggested earlier that we can imagine empirical findings that would lead us to revise the concept of hermaphroditism so that the statement "All hermaphroditic plants have both male and female reproductive parts" would become a contingent statement in need of test. Even if we accept this, it does not follow that the statement is *at present* testable. It is, at present, true by definition and so a test of it would be both impossible and irrelevant. Surely, this philosophical fact is one recognized by any NSF panel. In contrast, a statement such as "Hermaphrodites have, on average, higher levels of inbreeding than dioecious plants" is testable. Of course, I am not saying that all testable hypotheses should be tested; some are trivial and not worth the effort. The point is that *only* testable statements should be tested, that is, one should not try to test the untestable. For this reason the distinction between testable hypotheses and statements for which empirical testing is both impossible and irrelevant is important.

Is the PNS testable, does it have empirical content? On basis of the analysis of the PNS developed in chapter 1, this question is

[5] Note that if this is our concern, Mary Williams's "solution" to the tautology problem is a nonstarter. In her axiomatization of evolutionary theory she takes "fitness" to be an undefined term and the principle corresponding to the PNS to be an axiom (M. B. Williams 1970). Although it is legitimate to argue that because a certain principle need not be tested it may be taken as an axiom in some axiomatic system, it is wholly illegitimate to argue that because you have decided to treat a certain principle as an axiom in your system it need not be tested.

[6] Glymour (1980) has tried to work out a philosophical account of this sort of nonholistic hypothesis testing. For commentary on Glymour's work, see Earman (1983).

still not entirely straightforward. Recall that I argued for a propensity interpretation of relative adaptedness. According to that interpretation the adaptedness of an organism (in an environment) is a probabilistic disposition with some underlying causal basis. But, I argued, there is no single biological or physical property (no matter how complex) that underlies adaptedness for all organisms in all selective environments, because there is no such property that is invariably selected for. Thus I called for a schematic definition of relative adaptedness that correlatively makes the PNS schematic.

PNS: If *a* is better adapted than *b* in environment *E*, then (probably) *a* will have greater reproductive success than *b* in *E*.

Although when values for *a*, *b*, and *E* are specified it is possible empirically to test the resultant instantiation of the PNS, negative results of such tests do not invalidate the schematic "law" stated above. Thus, considered as a schematic "law," the PNS is not testable. As a schematic "law" the PNS allows us to infer inductively from the hypothesis that the expected reproductive success of *a* in environment *E* is greater than the expected reproductive success of *b* in *E*, to the conclusion or prediction that the actual reproductive success of *a* in *E* will be greater than that of *b*. Take out the term "reproductive success" and this inference is one we would make for any expected value. It is a part of probability theory called the *principle of direct inference*. It would be reasonable to argue that this principle is not testable, that it does not have any empirical content, but we need not concern ourselves with that issue here. Whatever one might say about that, it should be clear that the PNS has no empirical content of its own, that is, it has no biological empirical content. It is simply an application of probability theory to a biological problem.

When we instantiate our definition of relative adaptedness, that is, when we specify for a group of organisms in a given selective environment the causal bases of differential adaptedness, we get instances of the PNS. These instances are perfectly testable, but they lack generality, they apply only to a given group of organisms in a given selective environment. Recall from our discussion in section 1.6 that no amount of falsification of instances transfers up to the general schema itself. Thus the PNS as a general law is not testable; it has no empirical content (or put more cautiously, it has no biological empirical content).

## 4.2. THE ROLE OF THE PRINCIPLE OF NATURAL SELECTION

In this chapter I will argue that the PNS is the fundamental law or principle of the theory of natural selection. But if the PNS is simply an inductive inference rule, a rule that allows one to connect propensities to observed frequencies, then why consider it a part (indeed the central part) of a *biological* theory? Isn't it simply an application of a part of probability theory to a biological problem? Biologists, like other scientists, use inference rules from deductive and inductive logic. For instance, *modus ponens*, the inference rule that allows us to infer "Q" from "If P then Q" and "P," is frequently used by evolutionary biologists. Yet we do not feel compelled to treat *modus ponens* as a law of evolutionary biology. What is so special about the PNS? This is the question raised by Kenneth Waters (1986, in response to Brandon 1981b). Waters concludes that there is nothing special about the PNS and that it, like *modus ponens*, should not be considered a part of the theory of natural selection.

This line of reasoning seems compelling until one considers the role, or roles, the PNS plays in the theory of natural selection. The PNS as a general schema serves the role of systematic unification of the theory of natural selection, and it also serves as an organizing principle, a principle that structures natural-selectionist explanations.

### 4.2a. Systematic Unification

The standard story of the evolution of melanic forms of *Biston betularia* (and some other species) says that the darker-winged moths are better adapted to their environment (with dark tree trunks due to industrial pollution) than their lighter-winged conspecifics (Kettlewell 1973 is the classic source; for further references see Endler 1986, p. 134). Thus we get the following instantiation of the PNS:

If moth *a* has darker colored wings than *b* in (this specific) *E*, then (probably) *a* will have greater reproductive success than *b* in *E*.

Another classic case of natural selection involves the evolution of heavy-metal tolerance in plants growing on contaminated soils (Antonovics, Bradshaw, and Turner 1971). For these populations we get another instantiation of the PNS:

If plant *a* is more tolerant of heavy metals than *b* in (this specific) *E*, then (probably) *a* will have greater reproductive success than *b* in *E*.

What does heavy-metal tolerance have to do with dark-colored wings? Without the general schematic PNS we can only answer, "Not much." But with the general schema we can describe these two cases as two instances of the same phenomenon, namely, natural selection. That is, the PNS provides a systematic unification of these and other cases of natural selection. If all of natural selection were selection for heavy-metal tolerance, or if it were all selection for darker-colored wings, or energetic efficiency (as discussed in chapter 1), or whatever, then we would not need the general schematic PNS to unify all cases of natural selection. But as I argued in chapter 1, adaptedness is supervenient on biological properties such as heavy-metal tolerance, wing color, and energetic efficiency. Thus the general schematic PNS is necessary if we are to have a general theory of natural selection, as opposed to numerous unconnected low-level theories concerning the evolution of specific populations in specific selective environments.[7]

The virtues of systematic unification are probably too obvious to dwell on. It is arguable that the central business of science is systematic unification, that scientific explanation just is the unification of broad bodies of apparently diverse phenomena under some "law" or unifying generalization (see Friedman 1974 and Kitcher 1981). Of course, one might long for unifying generalizations that

[7] Beatty's (1981) approach is somewhat different. He questions the need for a general *principle* of natural selection. According to Beatty it suffices that there is a *definition* of natural selection that can be applied to particular cases (see pp. 411–413). This difference in emphasis may stem from his explicit adoption of the semantic view of theories (see 13 below), and does not represent a major disagreement.

are not schematic or, in other words, that are themselves testable. But as a physics professor of mine once said, "We didn't get here first." By that he simply meant that the world does not conform itself to our desire to do science in a certain way. If, as I have argued, the world is such that a general, testable, and empirically correct version of the PNS is impossible, then our best option is to adopt the general, but untestable, schematic PNS as a unifying principle. Of course, the instantiations of the PNS are perfectly testable, but they are specific to particular populations in particular selective environments.[8]

## 4.2b. Principle of Natural Selection as Organizing Principle

In chapter 2 I argued that natural selection is differential reproduction that is due to differential adaptedness to a *common selective environment.* I illustrated that point by the example of two different seed types distributed over two different selective environments. If one seed lands in nutrient-rich soil while the other lands in mildly toxic soil, then the first is likely to outreproduce the second. But this is not due to differences between the two seeds, but to differences in their environments. Thus we distinguish this case from cases of natural selection. Much of chapter 2 was devoted to arguing for the importance of this distinction. But if the PNS were *simply* an application of the principle of direct inference, then we would have no basis for drawing this distinction. A seed of a particular genotype in a certain soil type (under certain conditions of light, moisture, temperature, and so forth, which we will ignore) has a certain expected reproductive success that can be experimentally determined. A seed of a different genotype in a different soil type has its own expected reproductive success. By the principle

[8] The picture presented here—that of an untestable PNS at the core of the theory and its testable instantiations at the periphery—raises one question with which I will not deal. Are there generalizations about adaptedness that are more general than those that apply to particular populations but that are still specific and testable? I believe that there are and that such generalizations are of considerable biological interest, but their existence is largely irrelevant to the points I am trying to make. Sober (1984, pp. 51ff.) discusses Fisher's sex ratio theory is an example that would fall at this intermediate level of generalization, that is, an example that falls between the core and the skin.

of direct inference we can inductively infer differences in *actual* reproductive success between these two seed types in different soil types from differences in their *expected* reproductive success. From the point of view of the principle of direct inference, this case does not differ from one where the differences in expected reproductive success exist in a common selective environment. Probability theory is silent on which sort of comparison is of particular biological relevance.

But the theory of evolution by natural selection is not silent on this. Darwin's key insight was that under conditions of a "struggle for existence" organisms with slightly favored variant traits will (probably) outreproduce their competitors with less favored variant traits, and if these traits are heritable the distribution of trait values will change over generational time, with the favored variants becoming more common, that is, adaptive evolution will occur. The key point here is that the differences are among organisms, not among environments, and only differences among organisms stand a chance of being heritable. (The present focus is on organisms, but the point could be generalized to other levels of biological organization.) Why is this? If differences in expected fitness are due to differences in traits and the traits are heritable, then clearly differences in expected fitness can be heritable. But if differences in expected fitness are due to differences in environments, why can't these environmental differences be transmitted across generations so that again differences in expected fitness are heritable?

Environmental differences can be transmitted in two basic ways. The first is by the transmission of behavioral programs for habitat choice. But recall from our discussion in chapter 2, section 2.4, that habitat choice has the effect of damping out heterogeneity of the external environment so that organisms occupying quite different external environments are in fact occupying a common selective environment. What is heritable here are the behavioral differences, and selection is occurring in a common selective environment. That is, differences in external environments are transmitted, but this occurs within a common selective environment. The second way in which environmental differences can be transmitted is when a random process results in an unequal distribution that persists over a number of generations. For instance, a 100-year flood may unequally distribute two seed types over two different selective environments, and this unequal distribution may persist until the next big flood (although selection within selective environments may alter the distribution). Here there may be a positive covariance between seed type and selective environment over a number of generations, but that would not be counted as heritability of expected fitness because the differences could be totally accounted for in terms of environmental differences.

Thus the PNS has a powerful role in structuring natural-selectionist explanations. It directs us to look for differences in adaptedness to common selective environments. Only such differences can lead to natural selection and to adaptive evolution. A further example of how the PNS serves as an organizing principle in the theory of natural selection will be discussed in section 4.4, but the above considerations allow us to conclude that the PNS is not *simply* an instance of a law of probability theory.

I have discussed two roles of the PNS in the theory of natural selection: that of general schematic law, which allows for the systematic unification of the theory; and that of organizing principle, which structures explanations of evolution by natural selection. I have discussed these roles separately for expository purposes, but they seem to be two sides of the same coin. The PNS serves as an organizing principle because it is the generalization that unifies all cases of natural selection. But whether one separates these two roles or includes them together under one description makes no real difference. The point of this section is that the PNS is the central generalization of—is the core of—the theory of natural selection.

## 4. 3. The Empirical Presuppositions of the Applicability of the Principle of Natural Selection[9]

Thus far I have argued that the PNS is the core of the theory of evolution by natural selection. This principle is devoid of empirical biological content. At the periphery of the theory lie instantiations

---

[9] Parts of this section are taken with minor modifications from Brandon (1981b).

of the PNS; they are empirically testable hypotheses, but their empirical biological content is specific to particular populations in particular selective environments. Although I think this picture is correct insofar as it goes, it does present a puzzle. The puzzle is this: Is there no empirical biological content at the core of the theory of evolution by natural selection? Put another way, is the empirical biological content of the theory all specific and peripheral? That there are logical or mathematical elements at the core of an empirical theory should not be surprising (although this point seems to have escaped those critics of evolutionary theory who thought that the alleged tautologousness of "the survival of the fittest" invalidates the entire theory). On the other hand, it would be surprising if empirically empty statements formed the entire core of an empirical theory. What I have said so far may suggest that this is the state of the theory of natural selection. But it is not.

Although the PNS has no empirical biological content, the presuppositions of its applicability are empirical. (By "presupposition of applicability" I mean those conditions we presuppose when we try to apply the PNS.) The statement of these presuppositions forms the empirical biological core of the theory of natural selection. As I see it, this core consists of two basic claims and a third less central claim.

The first claim is that biological entities (some of them, at least) are chance set-ups with respect to reproduction. The term "chance set-up" is Ian Hacking's. According to him, "A *chance set-up* is a device or part of the world on which might be conducted one or more *trials*, experiments, or observations; each trial must have a unique *result* which is a member of a class of possible results" (Hacking 1965, p. 13). A biological entity in its environment is a chance set-up. The trial is the entity's life. The class of possible results is the class of possible numbers of offspring contributed to the next generation by the entity. Although to my knowledge Hacking is not explicit on this, for present purposes a set-up satisfying the above characterization counts as a chance set-up only if the various possible outcomes do have chances or probabilities. This is by no means mathematically necessary. Consider a coin and a tossing device. That is a set-up, but it is at least mathematically possible that it is not a chance set-up. On the limit relative frequency interpretation of probability, the probability of a result, say heads, is the limit of the relative frequency of that result in an infinite sequence of trials. Whether a sequence converges or not is not a mathematical necessity; indeed, we can imagine sequences where there is no limit. Consider, for example, the sequence 10 heads, $10^{10}$ tails, $10^{10^{10}}$ heads, . . . . The relative frequency of heads does not converge to a limit in this infinite sequence, so a coin and tossing device generating such a sequence would not be a chance set-up. Of course, I have argued that with respect to defining adaptedness the preferred interpretation of probability is the propensity interpretation. In that view, to attribute a chance or probability to a set-up is to attribute to it some empirically determinable property. Whether or not some set-up has this property is not for probability theory to say. For instance, suppose that under very tightly controlled conditions certain organisms displayed no stable frequencies of offspring numbers from which probabilities could be estimated. As far as we could tell, such organisms would not be chance set-ups with respect to reproduction. Thus our first claim, that (at least some) biological entities are chance set-ups with respect to reproduction, is an empirical biological claim.[10]

I have stated this presupposition in a philosophical language that may be foreign to biologists, but the idea itself is familiar to biologists who have tried to set out the conditions necessary for the occurrence of natural selection. Thus Henry Horn (1979, p. 51, after MacArthur and Connell 1967) states that the following is a necessary condition of natural selection: "Each variant [in a population] has a characteristic reproductive success, measured ultimately by the number of its offspring who breed in the next generation." According to John Endler's (1986, p. 4) definition of natural selection, for selection to occur it is necessary that there be "a consistent relationship between that trait [in which there is variation in a population] and mating ability, fertilizing ability, fertility, fecundity, and, or, survivorship." (Also recall my discussion of Darwin's three principles in section 1.2, which follows Lewontin 1968.) This list is by no means exhaustive (see Endler 1986, p.

---

[10] Although this claim is testable, in the actual practice of evolutionary biology it functions more as a working hypothesis (a hypothesis we assume to be true—in this case based on good evidence) than as a hypothesis to be tested.

4, for further references); indeed, the idea goes back to Darwin (1859). To say that organisms are chance set-ups with respect to reproduction is just to say that particular organisms (or particular types of organisms) in a given environment have a characteristic reproductive success.

The term "biological entity" is being used as a placeholder, thus the claim as stated is essentially open. The sorts of things about which it might be made, the sorts of things "biological entity" ranges over, were discussed in chapter 3. They include genes, chromosomes, parts of organisms, organisms, groups of organisms, and perhaps even higher-level entities (avatars or species?). The open nature of this claim gives the theory of natural selection a hierarchical structure and will be discussed in detail in section 4.4.

The second claim is this: there are cases where biological entities in a given homogeneous selective environment differ in their adaptedness to that environment. Note that this claim is not universal, it just says that such cases exist. It is to those and only those cases that the PNS applies. This claim is empirical; there are alternatives it rules out. One is that differences in realized fitness are entirely due to chance. This is the explanation offered by the neutrality, or non-Darwinian, theory of molecular evolution (discussed in chapter 1). Another is the sort of case briefly discussed in the previous section and explored in detail in chapter 2. These are cases where differences in realized fitness are due to environmental differences, not differences in the adaptedness of organisms to a common selective environment. Clearly there are situations where these sorts of explanations are appropriate, and this is entirely compatible with the second claim, which is not a universal claim. What is important is that when and where our second claim is true, these alternatives are ruled out.

To appreciate the significance of the second claim, consider an example that differs from those discussed above and in chapter 2. Let us suppose that in an interbreeding population of organisms there is a chromosome inversion on which there are two gene loci. The inversion has the effect of preventing recombination between these two loci. At one locus are two alleles $A$ and $a$, at the other are $B$ and $b$. The A-locus codes for eye color and the B-locus for coat color. The eyes of carriers of the *AA* genotype are darker than those with the heterozygote *Aa*, which in turn are darker than those with *aa*. Also, carriers of *BB* have darker-colored coats than carriers of *Bb*, who in turn have darker coats than carriers of *bb*. As a matter of historical accident, *A* is linked to *B* and *a* to *b*. Thus of the nine possible genotypes only three, *AABB*, *AaBb*, and *aabb*, occur.[11] Let us suppose that there is strong selection for darker coat color but that eye color is selectively neutral.

Consider the A-locus. There are differences in realized fitness between alleles *A* and *a*, that is, on average the *A*'s produce more copies of themselves than do the *a*'s. Does the PNS apply here? That is, can we explain the difference in realized fitness between *A* and *a* in terms of differences in their adaptedness to a common environment? A gene's environment includes, among other things, its surrounding genes. There is a crucial difference between the environments of *A*'s and *a*'s. The *A* genes always find themselves linked to *B* genes, the *a* genes are always linked to *b* genes. In studying the evolution at the A-locus we cannot abstract away from this difference in genetic environments, for it is precisely this difference that accounts for the fitness difference between *A* and *a*. Thus claim 2 is not applicable to them, and so the PNS is not applicable to them either. At the genic level this is not a case of differential reproduction due to differential adaptedness to a common selective environment but rather a case of differential reproduction due to differences in environments.

This does not imply that natural selection is not occurring at all; in fact, by hypothesis it is. It just is not occurring at the genic level, but at a higher level. (According to the screening-off analysis developed in chapter 3, it is occurring at the organismic level.)

This example of gene linkage is not too unrealistic. Gene linkage of this sort is a common enough phenomenon. It along with epistasis are the genic phenomena that have led some to argue that in general genes are not units of selection (Lewontin 1974; Wimsatt

[11] As George Williams has pointed out to me, if *A* and *B* are 100% linked, as in my example, then they would not count as separate genes in any of the standard formulations of population genetics. However, we can distinguish the two parts of the inversion on the basis of their phenotypic effects. Moreover, whether we call *A* and *B* genes or lengths of DNA is of no consequence with respect to the point of my example.

1981). This argument can be restated rather simply in terms of the present description of the structure of the theory of natural selection: In general, genes are not units of selection since in general the PNS is not applicable to them. (I will return to this point in the next section.) I have introduced the gene linkage example here to illustrate the empirical biological content of claim 2 and the power of the PNS as an organizing principle.

To summarize, the presuppositions of the applicability of the PNS form the empirical biological core of the theory of natural selection. I have argued that there are two basic presuppositions:

1. (At least some) biological entities are chance set-ups with respect to reproduction.
2. Some biological entities differ in their adaptedness to their common selectively homogeneous environment, this difference having its basis in differences in some traits of the entities.

When and where these presuppositions are satisfied, the PNS is applicable to the relevant entities. It will predict that natural selection will occur among these entities, that is, that these differences in adaptedness will result in differences in reproductive success.

These two claims form the empirical core of the *theory of natural selection*. But for natural selection to have an evolutionary effect, the differences in the traits that underlie adaptedness must be heritable. Thus to get the *theory of evolution by natural selection* we must add a third claim:

3. In some cases adaptedness is to a degree heritable, or, equivalently, the causal bases of adaptedness values are to a degree heritable.

This claim is less central than the first two since natural selection can occur without it being satisfied. Yet unless it is satisfied *evolution* will not occur by natural selection. Thus satisfaction of the first two claims is necessary for the occurrence of natural selection; satisfaction of the third is necessary for this natural selection to result in evolutionary change.

Note that the second claim implies the first. If some biological entities differ in their adaptedness to a common environment, they must be chance set-ups with respect to reproduction in that environment. Otherwise talk of their adaptedness to that environment would not make sense (recall the definition of relative adaptedness given in chapter 1). But clearly the first claim does not imply the second. Furthermore, the third claim implies the second. As discussed in section 4.2b differences in organisms (or more generally, in biological entities) are heritable, not differences in environments. (And heritability requires variation; see discussion in section 3.7.) But again the second claim does not imply the third. Thus (1) could be satisfied when both (2) and (3) are not, and (2) could be satisfied when (3) is not. But when and where (3) is true, (1) and (2) must also be satisfied.

Logically, the most striking feature of claims (1)–(3) is that they are *existential* claims.[12] Consider the first claim. Although I suspect that it is a basic fact about organisms and perhaps about the other relevant biological entities that they are chance set-ups with respect to reproduction, it would have no impact on the theory of natural selection if a few bizarre organisms were discovered that were not chance set-ups. Thus even though I know of no exceptions to claim (1), I have stated it as an existential claim (I have put in the parenthetical "at least some"). For reasons already discussed at length, it is much more obvious that the second claim needs to be existential rather than universal. This is because differences in realized fitness may be due solely to chance, or they may be due to environmental differences (e.g., the seed dispersal case or the case of gene linkage). Again it is clear that the third claim needs to be existential; even where there are differences in adaptedness to a common selective environment, these differences need not be heritable. Thus if I am right, the empirical core of the theory of natural selection, or more broadly the theory of evolution by natural selection, consists of existential claims. This flies in the face of the view of the structure of scientific theories that comes from the logical positivists (see Suppe 1977). According to that view, the empirical core of a scientific theory consists of empirical universal generalizations. The theory of evolution by natural selection has a universal generalization—the PNS—only it

12 "All crows are black" is a universal claim. Obviously, it is falsified by one non-black crow. "Some crows are black" is existential. It is compatible with the existence of nonblack crows, and only asserts the existence of black crows.

is not empirical; and it has an empirical core—claims (1)–(3)—only they are not universal.

If I am right, the positivist picture of the structure of scientific theories is surely wrong (at least in its application to evolutionary biology).[13] But this may be of little interest to those who are only marginally concerned with the history of ideas in philosophy of science. On the other hand, the differences between my view and the positivist view have important ramifications for our understanding of the theory of evolution by natural selection. From an empiricist's point of view, the major difference between universal and existential statements is in their susceptibility to falsification. We can easily falsify the universal "All crows are black." All we need is one nonblack crow. But how do we falsify the existential "Some fruit flies weigh two tons"? Radical Popperians deny that such statements are falsifiable, but that seems wrong if what is at issue is whether or not we can have good empirically based reasons to reject such a statement. Nonetheless, there clearly is an asymmetry between the two in their ease of falsification. Thus the empirical core of the theory of evolution by natural selection is not as susceptible to falsification as the positivists would have had it.

If existential statements are harder to falsify than universal statements, then the reverse is true with respect to verification. To verify "Some fruit flies weigh two tons" all we would need is a two-ton fruit fly. But verifying a universal generalization is obviously more difficult. Claims (1)–(3) do have positive instances, that is, there are cases where it has been demonstrated that natural selection is occurring on heritable variation (see Endler 1986). If an existential statement has a positive instance, then it is true; thus claims (1)–(3) are, *as a contingent matter of fact*, true. But that is just to say that the PNS has *some* applications and, of course, it says nothing about the relative importance of selection versus drift or

---

[13] A large body of literature exists in philosophy of science in general and in philosophy of biology in particular on the semantic conception of theories, a new rival to the old positivist view. My position in this chapter is, I believe, wholly compatible with the semantic view of theories. But discussion of this philosophical theory about scientific theories is not directly relevant to the points I want to make, thus I have not tried to translate my views into the language of the semantic conception of theories. Beatty (1981) was the first to argue that the semantic view best fits evolutionary theory. Perhaps the most complete discussion of the application of this view of scientific theories to evolutionary biology is to be found in Lloyd (1988).

any other candidate evolutionary force.[14] Thus the major question for the theory of natural selection is not whether natural selection occurs (we know that it does) but rather how frequent and how important is natural selection in the evolutionary process. And so critics of the theory should not hope to falsify it. The theory's central generalization, the PNS, is not falsifiable, and even though falsifying its empirical core would be possible, assuming that core were false, it would be very difficult. However, claims (1)–(3) are true, so falsification is out of the question. Critics would better spend their time arguing over the range of application of the theory. If that range were narrow, then the theory would not be falsified, but it would have only marginal value in evolutionary biology.

Of course some might take all of this to indicate the total bankruptcy of the theory of natural selection. But what's the complaint? Is it that true theories are difficult (indeed impossible) to falsify? That certainly does not seem to be a devastating criticism. Is it that the theory of natural selection does not fit a favored philosophical model of scientific theories? Again, I think this criticism is without force since I find it hard to justify terminating a viable and progressive research program simply because one of its basic theories fails to conform to some descriptively inadequate philosophical model. No, the real issue is not whether the theory is fundamentally flawed, nor whether it has *any* applications; the real issue is

[14] Gould and Lewontin (1979) argue that in biology most questions are questions of relative frequency or relative importance. Thus one does not ask whether all of evolution can be explained as the result of drift, or whether it all can be explained as the result of selection. Rather, the question is which is more important (in evolution in general, or, a more tractable question, in this particular evolutionary path). I should point out that questions of relative frequency and of relative importance are not the same. Suppose we had complete information, down to nucleotide sequences, of the evolutionary change in some group of organisms. Such information would tell us, for every change in the genome, whether it was due to selection or drift. We could then quantify the percent change due to these two factors. Suppose that at the nucleotide level 95% of the change was due to drift and only 5% due to selection. Would this mean that selection was relatively unimportant within this lineage? Not at all. Suppose that during the time interval in question the lineage had evolved some complex adaptation, say a wing that enabled the organisms to fly. Complex adaptations can evolve only by natural selection (see G. C. Williams 1966 or Dawkins 1987). So if our interest is in explaining complex adaptations, natural selection is *the* important factor to consider in this lineage, even though evolution by natural selection has been relatively infrequent.

the range of its application. That is to say, the real issue concerns the relative importance of natural selection in the evolutionary process.

## 4.4. THE HIERARCHICAL NATURE OF THE THEORY OF NATURAL SELECTION

Both Lewontin (1970) and M. B. Williams (1970) have argued that the abstract theory of natural selection could be applied to a number of levels of biological organization. I have also stated the PNS and its empirical presuppositions in ways that do not pick out any one level of organization. Thus the theory has a hierarchical structure. If the analysis of chapter 3 is correct, the theory can be applied to genes, chromosomes, parts of organisms, organisms, and groups of organisms. And this list is just a list of plausible levels at which the PNS can be applied; there may well be others. Now that I have described the abstract structure of the theory of natural selection, the question of what levels it applies to is left as an empirical question. I am by no means alone in holding this view;[15] indeed, I would almost be tempted to describe it as *the* accepted view in evolutionary biology were it not for the persistence and popularity of an alternative, explicitly nonhierarchical view. That view is called *genic selectionism*.

Genic selectionism has been advocated most strongly and ably by the biologists G. C. Williams (1966) and Richard Dawkins (1976), but also noteworthy are two papers by philosophers of biology that defend the position (Waters 1985; Sterelny and Kitcher 1988). The basic position is that all of natural selection can be accounted for (explained) in terms of selection for and against alternative alleles; that is, all of natural selection occurs at the level of individual genes.[16] There are two fatal flaws in this position. The

[15] A hierarchical view of selection is explicitly defended in the papers by Hamilton, Wimsatt, and Arnold and Fristrup in the Brandon and Burian (1984) anthology. Many other works either implicitly or explicitly adopt this view.

[16] For a very thorough treatment of these issues, see Lloyd (1988). She carefully distinguishes the positions of G. C. Williams (1966), the early Dawkins (1976), and the later Dawkins (1982b). It is doubtful that Williams really holds the position I have just set out, and it is not at all clear to me that the later Dawkins holds it. But these matters of exegesis are not my concern. I want to address only the position that, at the very least, has been inspired by these authors.

first is that it makes it impossible to distinguish adaptations of different level interactors. For example, what we might think of as a group adaptation and an organismic adaptation are indistinguishable because they would both be genic adaptations. All adaptations are genic adaptations according to the genic selectionist, and so the results of clearly distinguishable processes are conflated into one. This criticism will receive extensive discussion in chapter 5. But for now I want to concentrate on the second flaw, which is that the position requires a strange and perhaps incoherent notion of the environment within which selection takes place.

Sober and Lewontin (1982) have argued that genic selectionists cannot account for selection even in simple cases of heterozygote superiority. Consider a standard case of heterozygote superiority, the case of the sickle-cell hemoglobin allele. Homozygotes for the standard allele, call them *AA*, produce normal hemoglobin but are vulnerable to malaria in those populations in Africa (and elsewhere) where malaria is prevalent. Homozygotes for the sickling allele, call them *aa*, suffer anemia, which is usually fatal. But heterozygotes, the *Aa*'s, have only mild anemia and are resistant to malaria, making them superior in adaptedness to either homozygote. Usually this sort of case is described as one of differences in genotypic fitnesses; the *Aa* is fitter than both *AA* and *aa*, and this is seen as shorthand for a slightly more complicated phenotypic story. Sober and Lewontin (1982) argue that although one can algebraically derive *genic* fitness values from the *genotypic* fitness values plus genotypic frequencies, these genic fitness values are artifacts that obscure the actual causal process. In the language developed in chapters 2 and 3 we can say that sickle-cell whole genotypes, or more properly their resultant phenotypes, are interactors within a common selective environment, where malaria is a selective agent and where normally functioning hemoglobin is highly beneficial. Although I agree with Sober and Lewontin, G. C. Williams (1966, pp. 58–59) offers a genic selectionist account of just this case (see also Waters 1985 and Sterelny and Kitcher 1988).

Williams argues that in order to assign selection coefficients to genes meaningfully, we must take into account their *genetic environment*. In the case of heterozygote superiority, the fitness of a particular copy of a particular allele depends on what else is present at that locus. Thus a particular copy of the *a* allele will have very low, if not zero, fitness if it finds itself paired with another *a*; but if it is paired with an *A* it has high fitness. The overall fitness of *a* is the sum of its fitness in the different genetic environments weighted by the relative frequency of those environments. In the general case, the fitness $\overline{W}$ of a gene can be defined as follows:

$$\overline{W} = \Sigma\ (P_iW_i),$$

"where $W_i$ is the coefficient in the *i*th genetic environment, and $P_i$ is the relative frequency of that environment" (G. C. Williams 1966, p. 59). In our simple case of one locus and two alleles, *a* can find itself in only two genetic environments, next to another *a* or next to *A*. Thus the fitness of *a*, $\overline{W}_a$, is given as follows:

$$\overline{W}_a = 2P_{aa}\ W_{aa} + P_{Aa}W_{Aa},$$

where $P_{aa}$ is the relative frequency of *aa* and $W_{aa}$ is the fitness of the genotype *aa*. (Note that $P_{aa}W_{aa}$ is counted twice. This is because for *a* the relative frequency of the genetic environment where *a* is the other allele is $2P_{aa}$.) Similar remarks apply to *A*. Although the expected fitness of an allele is *context sensitive* (Sober and Lewontin's term), this is not an obvious reason to exclude assigning fitnesses to genes, since the adaptedness of a biological entity is always relativized to a given environment (chapter 1).

Williams distinguishes the *genetic environment* from the *somatic environment*, and that from the *ecological environment* (see 1966, pp. 61ff). We have just discussed the genetic environment. Although Williams admits that the distinction between it and the somatic environment will sometimes be difficult and arbitrary, the basic idea is that the somatic environment mediates and influences the expression of the genome. For instance, in cases of cytoplasmic inheritance genotypically identical organisms could develop in quite different ways (and these differences could affect relative fitness). Williams's notion of ecological environment is close to, if not identical with, the concept of external environment developed in chapter 2. Is Williams suggesting that the difference between the standard account of heterozygote superiority and the genic account is over the sort of environment to which fitnesses are made relative? I believe he is, and this is clearly the suggestion of Waters

(1985) and Sterelny and Kitcher (1988). Waters argues that neither account is right nor wrong, that both are equally acceptable, and that only pragmatic factors could lead one to choose one over the other. Sterelny and Kitcher and, I believe, Williams suggest that the genic account is inherently superior because it is more general. It is more general because summing over all genetic environments implicitly includes the relevant somatic and ecological factors, but not the other way around. We cannot say what the fitness of a particular copy of allele *a* is in a given somatic and ecological context without knowing its genetic environment. But when we actually *measure* the fitness of *a* in a given genetic context, we are doing so in some somatic and ecological context. So when we sum over all genetic contexts we are also summing over all the relevant somatic and ecological contexts.

I must admit that although I think that this argument lies behind the claim that the genic selectionist account is more general than and so superior to other accounts, I do not find it directly stated in any of the aforementioned works. Nonetheless, I think that it is worth criticizing. In my view the most fundamental problem with the above argument is that selection (in the cases under consideration) *does not occur in genetic environments*. Some selection processes do occur within genetic environments. In chapter 3 we discussed cases of "selfish" DNA (*sensu* Orgel and Crick) and cases of meiotic drive, both of which are selection processes that occur *within* genetic environments. But these processes are not at issue here; we are at present concerned with standard cases of what is normally thought of as organismic selection. Let's return to the case of heterozygote superiority. There are two alleles in the heterozygote and, barring "jumping" genes or meiotic drive, their fate is tied together. If the organism in which they are housed has high reproductive success, they both benefit equally (ignoring probabilistic noise); if the organism dies before reproducing, they again are affected equally. There simply is no selection between the two alleles in that context.

In chapter 2 I argued that in explaining natural selection the relevant concept of environment is that of the selective environment. The selective environment is the arena within which selection occurs. In cases of organismic selection the genetic environment cannot be a selective environment if, as I have just argued, selection does not occur within it. But perhaps in taking the genetic environment to be a spatiotemporally localized region within an individual organism, I am construing the notion too narrowly. Perhaps in the one-locus, two-allele case we should think of the genetic environment as follows: one genetic environment is where *A* is the *other* allele of the locus; the other genetic environment is where *a* is the other allele. Thus *A* and *a* can find themselves in one or another of these two genetic environments. These "environments" are spatiotemporally discontinuous and intermixed. Moreover, one and the same spatiotemporal region can be two different environments. Thus in a heterozygote the *A* allele is in one genetic environment while the *a* is in the other, even though the two alleles are locked together in the same organism and will share a common fate. I think the best that can be said for this view is that it is not logically contradictory.

Dawkins (1982b) suggests that the genic selectionist account is an alternative way of thinking about natural selection as it is commonly understood. He suggests that by translating our more common descriptions of natural selection into genic selectionist terms we gain generality and simplicity and lose nothing. But the account of selective environments offered in chapter 2 gives a framework for genuine ecological explanations of differential reproduction. I see nothing comparable on the genic selectionist side. And the account of interactors given in chapter 3 is perfectly general as it allows for the possibility of selection at any level of biological organization. The defenders of genic selectionism have yet to show that their account can produce anything other than post hoc redescriptions of selection scenarios that have already been developed within the framework defended herein. They have yet to show that their view has any explanatory or predictive power.

These topics will be discussed further in the next chapter, where we will see that as a theory of adaptation the genic selection theory is not a serious rival at all. For now the important point is that the theory of natural selection, as I have described it, does have a hierarchical structure. The theory itself does not determine which level or levels of biological organization fall under its domain. That is left as an open empirical question.

## 4.5. SUMMARY

I have argued that the central generalization of the theory of evolution by natural selection is the PNS. It is a schematic law without empirical biological content. It is the generalization that allows for the systematic unification of the theory of natural selection. Without it we would only have various unconnected low-level theories about selection in particular populations in particular selective environments. It is also an organizing principle that structures natural-selectionist explanations. Although it lacks empirical content, the presuppositions of its applicability are empirical. The two most basic presuppositions are that biological entities (at least some of them) are chance set-ups with respect to reproduction, and that in some cases biological entities differ in their adaptedness to a common homogeneous selective environment. If these two assumptions are satisfied, then the PNS is applicable. A third slightly less central presupposition is that in some cases adaptedness is to a degree heritable. When the third is true, natural selection can have evolutionary consequences. All three of these presuppositions are existential; they are not lawlike in the traditional positivistic sense. This has important consesquences for the testing of the general theory.

The PNS and its three presuppositions form the core of the theory of evolution by natural selection. This core says nothing about which particular level or levels of biological organization satisfy it. Thus the theory is potentially hierarchical.

At the periphery of the theory are the instantiations of the PNS. These instantiations have to do with particular populations in particular environmental settings. They are straightforwardly testable but lack generality. Exploring these instantiations comprises the difficult and important empirical work that still needs to be done on the theory of natural selection.

# *Mechanism and Teleology*

This final chapter explores the nature and value of adaptation explanations. What is required for a complete adaptation explanation? In answer to this question I present an account of what I call *ideally complete adaptation explanations*. Due to epistemological limitations this ideal is rarely, if ever, realized in evolutionary biology. But I argue that adaptation explanations that fall short of the ideal in certain ways are still of considerable value. Are adaptation explanations teleological? I argue that they are in an important sense, but I also show how this teleological aspect must be explicated in purely mechanistic terms. Finally, I argue that genic selectionism, or any single-level theory of adaptation, is explanatorily inadequate in a world in which selection occurs at multiple levels.

## 5.1. Two Concepts of Explanation

Before I explicate the structure of adaptation explanations in the theory of evolution by natural selection, it might be useful to describe briefly the nature of scientific explanations in general. Unfortunately there is presently no consensus among philosophers of science on this topic. During the 1950s and early '60s a consensus did form around the views of Carl Hempel. But since the mid 60s this consensus has succumbed to numerous critiques, and now philosophers of science seem to agree only on the inadequacy of Hempel's theory.[1]

But all is not chaos. Philip Kitcher (1985 and 1989) and Wesley Salmon (1989) have suggested that there are two viable alternatives among philosophical theories of explanation.[2] One is the

[1] For the status of the Hempelian approach as of 1963, see Scheffler (1963). Salmon (1989) offers a comprehensive history of the theory of explanation since 1948.

[2] Salmon (1984 and 1989) argues that three basic conceptions of explanation have been extant since the time of Aristotle. They are the ontic, the epistemic, and the

view that scientific explanation consists in the *unification* of broad bodies of phenomena under a minimal number of generalizations (Friedman 1974; Kitcher 1981, 1989). According to this view, the (or perhaps, a) goal of science is to construct an economical framework of laws or generalizations that are capable of subsuming all observable phenomena. Scientific explanations organize and systematize our knowledge of the empirical world; the more economical the systematization, the deeper our understanding of what is explained. The other view is the *causal/mechanical* approach (Railton 1978, 1981; Humphreys 1981, 1983; Salmon 1984, 1989). According to it, a scientific explanation of a phenomenon consists of uncovering the mechanisms that produced the phenomenon of interest. This view sees the explanation of individual events as primary, with the explanation of generalizations flowing from them. That is, the explanation of scientific generalizations comes from the causal mechanisms that produce the regularities.[3] In Kitcher's (1985) terminology the causal/mechanical approach is *bottom-up*; the unification approach is, on the other hand, *top-down*.

To illustrate the difference between these two approaches, Salmon (1989) tells a story about "the friendly physicist." As the story goes, a friendly physicist was on a jet airplane waiting to take off. Across the aisle was a young boy holding a helium-filled balloon by a string. The physicist asked the boy what would happen to the balloon on takeoff. The boy replied that he thought the balloon would move toward the rear of the cabin. This seemed intuitively right to the adults overhearing the conversation. But the physicist said that he thought it would move forward. When the plane accelerated to takeoff the balloon moved forward, much to the boy's surprise.

How do we explain this phenomenon? Salmon offers two explanations. The first is a unification explanation. According to the theory of general relativity (the principle of equivalence), an accel-

---

modal. Neither Salmon nor I are sympathetic toward the modal conception. The causal/mechanical approach discussed later in this section is a version of the ontic conception, and the unification approach is a version of the epistemic conception.

[3] It is not entirely clear just how this works; indeed, this is the outstanding problem for the causal/mechanical approach. But for our purposes a critical discussion of this and related issues is not necessary. These issues are carefully evaluated in Salmon (1989) and Kitcher (1989).

erating reference frame is equivalent to a reference frame in a grav-itational field. On the earth's surface we can see that a helium bal-loon tends to move away from the source of the gravitational field (i.e., it rises). Likewise, in an accelerating plane the helium balloon moves away from the *equivalent* of the source of the gravitational field, that is, it moves forward. The second explanation is causal/mechanical. When the rear wall of the cabin moves forward on acceleration, it strikes many nearby molecules of the air in the cabin, thus creating a pressure gradient from the back to front of the cabin. Because of that, more molecules strike the rear of the balloon than the front, and so the balloon moves forward.

Which is *the* correct explanation? Salmon argues, and I agree, that neither is *the* correct explanation; rather, both are perfectly le-gitimate. The first increases our understanding of the phenome-non by subsuming it under a very general physical principle. The second increases our understanding by describing the relevant un-derlying mechanisms.

What is the relationship between unification and causal/me-chanical explanations? Three possibilities suggest themselves: (1) The two approaches offer distinct but compatible, indeed comple-mentary, explanations of some given phenomenon; (2) the two ap-proaches represent two sides of the same coin; they do not corre-spond to two distinct but complementary explanations, but a complete explanation of some phenomena has both unification and causal/mechanical aspects; and (3) the two approaches are in-compatible rival theories of explanation. Salmon's discussion of the friendly physicist seems to suggest (1), though later (p. 390) he seems to endorse (2). Kitcher (1985, 1989), on the other hand, seems to be committed to (3). I will not try to address this question in its general form, but in section 5.4 I will argue that alternative (2) captures some important features of natural selectionist expla-nations of adaptations.

## 5.2. ADAPTATION EXPLANATIONS

When philosophers present a theory of explanation they usually present a theory of what might be called *ideally complete explana-tions*. If, as is the case, ideally complete explanations are rarely if ever given in the practice of science, one might think that this is a typical case of philosophy of science not being sensitive to the actual practice of science and so being irrelevant for it. But such a thought may not be justified; it depends on why ideally complete explanations are not prevalent within the practice of a particular branch of science. There are two reasons why the *ideal explanatory text* might not be filled in.[4] First, parts of it, or perhaps all of it, may be impossible to fill in. For instance, suppose that according to my theory of explanation in psychology, ideally complete psychological explanations are to be given in quantum-mechanical terms. Such a theory would, quite rightly, be ignored. The second reason is that the text as a whole may be practically impossible to fill in because of the time and effort that would be required, but no *part* of it presents, in principle, difficulties.[5] In this case a theory of ideally complete explanations can be a useful model against which actually proffered explanations can be compared. When putting forward an incomplete explanation, one would like to know the ways in which it is incomplete. If the point of scientific explanation is to yield scientific understanding of the explained phenomena, then knowing the potential pitfalls of proffered explanations and knowing what would count as a complete explanation clearly enhance our understanding of the phenomena and are therefore of explanatory value.

In this section I will present an account of natural-selectionist explanations of adaptations. Although, as discussed in chapter 1, there are other possible explanations of organic adaptations, such as the creationist or Lamarckian accounts, there is good reason to believe that the natural-selectionist account is the only scientifically viable one. So when I speak of adaptation explanations I mean natural-selectionist explanations of adaptations. In my attempt to present the necessary components of an ideally complete adaptation explanation, I may inadvertently leave out some types

[4] The term *ideal explanatory text* is Railton's (1980). He distinguishes between the ideal explanatory text and what he calls *explanatory information* which is explanatory but falls short of the ideal. Similarly, Hempel (1965) distinguishes between genuine explanations and explanation sketches.

[5] This is not very precise. The idea is that the whole explanation may be so detailed and complicated that it will never be produced, but that any part of it could be produced. Of course, that would depend on how the whole was broken up into parts. I assume that the parts are manageable in size. When we get to the detailed discussion of adaptation explanations, I think this will become sufficiently clear.

of information that are explanatorily relevant in some cases. Perhaps I should therefore say that I am offering a minimal account of ideally complete adaptation explanations, but in what follows I will ignore this caveat.

In chapter 1 (section 1.8) I argued that adaptation is a historical concept, that adaptations are the direct products of the process of evolution by natural selection. So, to say that something is an adaptation is to make a claim about its causal history. In chapter 3 (section 3.2) I argued that with respect to natural selection, phenotypes screen off genotypes, or more generally, interactors screen off replicators, and so adaptations are features of phenotypes or interactors. The process of adaptation is simply the process of evolution by natural selection (section 1.8). I will now argue that to explain an adaptation is to show how and why the process of adaptation produced its product.

When we think of adaptations we typically think of "big" adaptations, complex features such as mammalian eyes or avian wings, features Darwin called "organs of extreme perfection." The process that produces complex and intricate adaptations presumably takes place over long periods of time and runs through lineages that have branched through many speciation events. At the other extreme are cases of adaptation that take place within a species over relatively short periods of time. We will return to the "organs of extreme perfection" in the next section, but for now I will concentrate on examples that represent the simplest cases of adaptation.

Perhaps the best-studied examples of the process of adaptation taking place within subpopulations of a species (the evolution of races or ecotypes) are cases of the evolution of heavy-metal tolerance in plants. This has occurred in many different species in quite different settings, for example, at the edges of mines, at ore smelting plants, along roadsides (polluted by lead), and on naturally occurring outcrops that are rich in heavy metals (Antonovics et al. 1971).[6] For instance, along the edges of mines, soil contaminated with one or more heavy metals can form very sharp boundaries with noncontaminated soil. Grasses (e.g., *Agrostis tenuis, Festuca*

[6] Antonovics, Bradshaw, and Turner (1971) is a review article on heavy-metal tolerance in plants. The information that follows comes from this article.

163

*ovina*, or *Anthoxanthum odoratum*) growing on the contaminated soil are much more tolerant of the underlying metals than are their conspecifics growing a few feet away from them. This difference in tolerance has a genetic basis, thus there is genetic differentiation of the population in spite of considerable gene flow (in the form of seeds and, especially, pollen) between the contaminated and non-contaminated sites. A number of lines of evidence indicate that the tolerant forms have evolved from the nontolerant, and have done so many times in different species and at different sites. Finally, overwhelming evidence points to very strong selection against the nontolerant types on contaminated soil (and also to strong, but not quite so strong, selection against tolerant types on normal soil). Thus the tolerant types are better adapted to the environment of contaminated soils than the nontolerant types, and since we have every reason to believe that tolerance has evolved in the subpopulation located on the contaminated soil in response to this differential adaptedness, heavy-metal tolerance is an adaptation in this subpopulation.

How do we explain this adaptation? First, we must clarify exactly what we are trying to explain. If we take some particular plant growing on the contaminated soil and ask why it is metal tolerant, a natural answer would be that it developed from a seed of such-and-such genotype and seeds of that genotype grown in that environment result in plants that are tolerant of heavy metals. That is, we might start with the seed (with its particular genotype and cytoplasm and whatever else might be relevant) and give a developmental answer to the question. Such an explanation is perfectly legitimate, but does nothing to explain the adaptation, that is, it sheds no light on why tolerant plants are common on contaminated soils.[7] What is required is an evolutionary explanation. Individual organisms do not evolve in response to natural selection;[8] they die or they thrive but they do not evolve. Populations—

[7] Not that it couldn't. If metal tolerance were an environmentally induced trait, then the developmental answer may be just the one we are looking for. But in that case heavy-metal tolerance would not be an adaptation in the strict sense. Perhaps the phenotypic plasticity that allows for the development of heavy-metal tolerance in contaminated environments would be an adaptation, but that is another matter.

[8] More precisely, they do not evolve in response to organismic selection. They may respond to suborganismic levels of selection, for example, developmental or somatic selection, as discussed in chapter 3.

and species and clades, but for the moment we will concentrate on populations—evolve in response to natural selection. In the first instance then, evolution can explain features of populations, but it can only derivatively explain features of the component individuals. In our case we may offer an evolutionary explanation of why all, or nearly all, of the plants growing on the contaminated soil are heavy-metal tolerant, which derivatively allows us to explain why this *particular* plant is heavy-metal tolerant.

## 5.2a. Five Components of Complete Adaptation Explanations

I will argue that an ideally complete adaptation explanation requires five kinds of information: (1) Evidence that selection has occurred, that is, that some types are better adapted than others in the relevant selective environment (and that this has resulted in differential reproduction); (2) an ecological explanation of the fact that some types are better adapted than others; (3) evidence that the traits in question are heritable; (4) information about the structure of the population from both a genetic and a selective point of view, that is, information about patterns of gene flow and patterns of selective environments; and (5) phylogenetic information concerning what has evolved from what, that is, which character states are primitive and which are derived. We will go through the explanation of the evolution of heavy-metal tolerance to illustrate the roles of each of these categories. But first I must stress that their ordering is of no significance. In particular, the list is not intended to be an algorithm for constructing adaptation explanations. One might start work on an adaptation explanation at any point in this list, and go through it in any order; the important thing is that for an adaptation explanation to be complete it must contain information from each of the five categories.

### SELECTION HAS OCCURRED

Clearly, if an adaptation is the product of evolution by natural selection, selection is a prerequisite for adaptation. Establishing that selection has occurred in the relevant population(s) over the relevant time interval can be problematic. There are two basic reasons for this. The first is that if selection has already driven the

165

selected trait to fixation, then we cannot observe selection in action. Sometimes this problem can be circumvented by experimentally introducing the relevant variants (assuming we know what they are). Otherwise we have to rely on indirect historical evidence, such as a fossil record indicating that one type has replaced another. (This sort of evidence is indirect, since it indicates only a direction of change; it does not necessarily implicate selection.) The second is that whether we observe selection among experimentally introduced or naturally occurring variants, we have to extrapolate back through time and claim that the selective environment we are presently observing is similar in its relevant aspects to past selective environments.

Although these epistemological problems can be critical and will receive further discussion in the next section, they are not problematic with respect to the evolution of heavy-metal tolerance. First, gene flow constantly brings nontolerant individuals into the contaminated area, so that selection can be observed during any year. Indeed, the observed selection pressures against the nontolerant types are quite intense (from 0.3 to 0.7 and perhaps higher; Antonovics et al. 1971, p. 39). Second, in some cases we know that the evolution of tolerance has been rapid and recent (e.g., on mine sites that are only 50–100 years old), so we can confidently extrapolate back from the current situation. Thus in this case we have good reasons to believe that strong selection has continually occurred in the relevant populations.

ECOLOGICAL EXPLANATION OF SELECTION

It is one thing to know that selection has occurred, it is another thing to know *why* it has occurred. That is, we may know that one type is better adapted than another to a given selective environment but not understand the ecological reasons for this (see discussions in sections 1.8 and 2.6). To return to the example (from Lewontin 1978) discussed in chapter 1, we may observe that darker-colored organisms are better adapted than lighter-colored ones in some specific environment. This may be due to differential predation, as in the case of melanic moths in polluted woods, or it may have nothing to do with coloration. It may be that some organisms produce an enzyme that helps to detoxify a poison common in the environment, and in the process of detoxification the enzyme is converted into an insoluble pigment. In the first case, the selective agent is predation and the adaptation is crypsis. In the second case, the selective agent is a toxin in the environment and the adaptation is the ability to cope with that toxin. A large measure of the force of Gould and Lewontin's (1979) critique of the "adaptationist programme" comes from the problems associated with giving ecological explanations of selection. Oftentimes "just so" stories are substituted for difficult ecological analyses. Of course, the problem of producing ecological analyses of observable selection regimes is compounded with the above-mentioned problem of extrapolation when we offer these ecological explanations for past selective events.

But again, these are not problems in the case of the evolution of metal tolerance. Here the relevant ecological differences are obvious. The different concentrations of metals in contaminated vs. uncontaminated soils are easily measured, and the effect of soil type on the different types of plants is also easily detected. Tolerant types outperform nontolerant types on contaminated soil. (And nontolerant types outcompete tolerant types on noncontaminated soil, but that is not important to our discussion here.) This is about as straightforward as an ecological explanation of selection can be.

One might argue that a full understanding of the evolution of metal tolerance would involve more than the ecological explanation of selection; it would also include a physiological or biomechanical explanation of the mechanisms of metal tolerance in the relevant plants. Interesting as this work is, I do not consider it a necessary part of the evolutionary explanation of the adaptation. Whereas the ecological explanation tells us *why* one type is better adapted than another in a given environment, the physiological/biomechanical account tells us *how* one type manages to be better adapted than the other. It answers a different question.[9]

---

[9] One can easily imagine cases where the physiological or biomechanical account will be relevant to a complete adaptation explanation. For instance, if the physiological/biomechanical account implies a certain limitation in the relevant adaptation (e.g., a limit to the level of heavy-metal tolerance) or a necessary trade-off in achieving the adaptation (e.g., metal tolerance being inversely related to fecundity), then such information will play a role in a complete adaptation explanation. My point is simply that selection *can* be indifferent to the underlying physiology or biomechan-

167

### HERITABILITY OF TRAIT

For natural selection to have evolutionary consequences the relevant variation must be heritable. Thus if we are to explain the presence of a certain trait in a population by means of an adaptation explanation, that trait must have been heritable. Although this component of a complete adaptation explanation presents less problems than our other categories, it still presents some difficulties.

Obviously, selection implies variation. It is equally true, but less obvious, that heritability implies variation (see discussion in section 3.7). Thus if selection has driven the relevant trait to fixation in the population, we cannot, without experimental manipulation, observe selection on the trait; likewise, we cannot observe the degree to which the trait is heritable. For example, if all *Anthoxanthum odoratum* are equally metal tolerant, the heritability of that trait will be zero in that population. Since most of the "big" adaptations that people want to explain are fixed in the species or higher taxon, this problem is quite general.[10] But I do not view it as a very serious problem. The indirect evidence that selection has occurred in the past (discussed above) can also indirectly support the heritability of the trait. If we have historical evidence that trait $A'$ has replaced $A$ in some lineage, it weakly supports the claim that trait values in $A$ were heritable. The support is only weak since that same pattern can be produced in a different manner, for example, if $A'$ is environmentally induced and the environment has gradually changed (in the right way) through time, then $A'$ would gradually replace $A$ in the lineage. This evidence can be strengthened if the trait is the sort of trait that is typically heritable in closely related groups.[11]

---

ics, and so I do not include such information in my general account of the essentials in a complete adaptation explanation.

[10] Perhaps the problem is not so general after all. If one thinks of discrete valued traits, such as eyes or no eyes, then it does seem general. But if we think of the same traits in quantitative terms, then it does not. For instance, presumably there is quantitative variation in eye characteristics in many vertebrate populations. How else could eyeless vertebrate species evolve in caves?

[11] I realize this is vague, but I think it is clear enough for present purposes. In humans, for example, hair color is the right sort of trait, hair style is not. As pointed out above, because $A'$ has replaced $A$ in a lineage does not necessarily

A proffered adaptation explanation may fall short of the ideal with respect to the heritability of the relevant trait in another way. At minimum we need to know that the trait is heritable, but a complete account of the evolution of an adaptation makes reference to the underlying genetics of the trait. The reason for this is straightforward: the underlying genetics affects the evolution of the trait in the population and our adaptation explanation is given in terms of this evolutionary trajectory. To give a simple example, suppose metal tolerance is controlled by a single gene with two alleles. In that case, whether one of the homozygotes or the heterozygote is most tolerant will make a difference to the evolutionary response to selection. Let's say this in a slightly different way: a quantitative genetic measurement of the heritability of some trait in some particular population in some particular environment is an empirical constant which may or may not remain constant as the population and/or environment change. To reliably extrapolate heritability values, we need to know the underlying genetics.

In the case of the evolution of heavy-metal tolerance the first potential problem does not arise. There is variation in metal tolerance, so its heritability can be readily measured. In all cases metal tolerance has proven to be heritable. But the underlying genetics is not entirely known.[12] Here is one area where the present account of the evolution of metal tolerance is incomplete.

POPULATION STRUCTURE

Much theoretical work in population biology has been devoted to exploring the evolutionary consequences of different population structures. Consider, for instance, Sewall Wright's famous shifting balance model of evolution. Models of kin and group selection are similarly directed. But population structure has two independent aspects. The first is what might be called the *demic structure* of a population. This structure reflects the patterns of gene flow. The second is what might be called the *structure of the selective environ-*

---

mean that there has been selection for $A'$; $A$ and $A'$ may be selectively neutral and $A'$ may have drifted to fixation.

[12] Because metal tolerance has evolved in many different species at many different sites, we should not expect a uniform genetic story. Investigators of some cases have uncovered some preliminary information about the underlying genetics—for instance, that metal tolerance is a polygenetic trait. See Antonovics, Bradshaw, and Turner (1971, section G).

*ment*, which reflects patterns of selective heterogeneity and homogeneity (discussed in detail in chapter 2). Models of group selection explicitly involve both aspects of population structure: in interdemic models, groups are demes; in intrademic models, group members disperse into a common mating pool—the deme. In both sorts of models, variation between groups makes each group a selective environment (for the contained individuals). Thus in the interdemic models demes are selectively homogeneous; in the intrademic models demes are selectively heterogeneous (see chapter 3).

In empirical studies the demic structure of the population is more likely to be attended to than is the structure of the selective environment. There are two reasons for this. First, the demic structure is usually easier to ascertain. Second, I believe that there is a tacit assumption among many biologists that demes are selectively homogeneous. Such an assumption would lead to the neglect of patterns of the selective environment. But, as I argued in chapter 2, these patterns of differential reproduction within demes may or may not accurately reflect selection. (Recall the example in section 2.3 of the two seed types unequally dispersed over two selective environments.)

There can be no doubt that different population structures make an evolutionary difference and so are a part of a complete adaptation explanation. It is equally obvious that historical knowledge of past population structures will usually be problematic. The problems are not different in kind from those discussed above. We need some basis for extrapolating observational or experimental data on present populations to the relevant past populations. Sometimes the extrapolation is easily justified. For instance, if we are concerned with the evolution of wings in insects and our hypothesis is that they are adaptations for flight, we can reasonably assume that the laws of aerodynamics have held constant throughout the relevant time period. Thus current experiments are relevant to the past selective environment. But explaining current patterns of human selfishness and altruism in terms of population structure of early hominids is a bit more problematic.

One of the reasons I chose the example of metal tolerance in plants is that it is the example of evolution by natural selection where the interaction of gene flow and the pattern of selective environmental heterogeneity is most important and best understood. Prior to empirical investigation of these cases, it was thought that natural selection could not sharply differentiate populations in the presence of high levels of gene flow. Recall that there is selection for tolerance on the contaminated soil and selection against tolerance on the surrounding noncontaminated pastureland. Also, pollen (and presumably some seed) is carried from one site to the other. Antonovics et al. (1971, p. 40) show how the interaction between gene flow and selection is beautifully illustrated by the

> discovery of a copper mine in a U-shaped valley with the prevailing winds in one direction, where a study was made of transects across the mine boundaries in a cross-wind and down-wind direction (McNeilly, 1968). The cross-wind transect showed a sharp cline at the mine boundary whereas in the down-wind transect the tolerance decreased gradually over a long distance (100m) into the pasture. . . . [This] shows clearly the interaction between selection and different amounts of gene flow in determining the pattern of differentiation.

Here we can directly observe both processes—gene flow and selection—that have produced the patterns we seek to explain.[13]

## PHYLOGENETIC INFORMATION ON TRAIT POLARITY

Although I have listed this category last, perhaps it should be listed first, since without information about what has evolved from what, we cannot meaningfully hypothesize that some trait is an adaptation. Adaptations evolve because their possessors are better adapted to some particular selective environment than are posses-

[13] One of the questions I raised at the end of chapter 2 is whether or not we should expect selection to bring genetic neighborhoods more nearly in line with selective neighborhoods when gene flow across selective environments slows or prevents total adaptation to the local selective environment. There is some evidence to suggest that this is occurring in some populations. Differences in flowering time between the tolerant and nontolerant populations have been observed, and the self-fertility of tolerant types has been shown to be greater than that of nontolerant types. Here selection seems to be changing the breeding system to impede gene flow between the two selective environments. See Antonovics, Bradshaw, and Turner (1971, p. 41, and references therein).

sors of alternative traits. Metal tolerance evolves in plant populations growing on contaminated soils because tolerant types are
better adapted to that environment than nontolerant types. But
suppose that the members of some plant species were all metal
tolerant prior to that species (or its ancestors) ever having been
exposed to contaminated soil. In that case, the fact that the plants
growing on contaminated soil were metal tolerant would not call
for an adaptation explanation. Metal tolerance would not be an
adaptation in this species, it would be an *aptation*. This term was
coined by Gould and Vrba (1982) to apply to traits that are used in
ways that increase the adaptedness of their possessors, but that
did not evolve because of that use.[14] The human nose is not an
adaptation to hold spectacles but it has been co-opted for that use,
and so (assuming there are fitness benefits from being able to see)
it is an aptation for that use. Similarly, in our hypothetical example
metal tolerance would be an aptation.

The necessity of information on what has evolved from what,
called *trait polarity*, is particularly evident at the supraspecific level.
J. A. Coddington (1988) gives six examples illustrating the pitfalls
of either ignoring trait polarity or getting it wrong. One example
from Coddington's own work (pp. 9–10) will suffice to make the
point:

A century-old tradition in arachnology asserts that the orb
web evolved from something like the cob web because it is a

[14] Gould and Vrba (1982) also coined the term "exaptation" which, together with
the term "aptation," replaces the older term "preadaptation." They objected to the
connotation of preadaptation, thinking that it suggested foresight or planning in
the evolutionary process. While I tend to agree with them, we should note that
some biologists object to the coining of these new terms when a perfectly acceptable one already has an established usage. By citing Gould and Vrba (1982) I do not
mean to suggest that earlier biologists had failed to make this distinction, whatever
we may call it. Gould and Vrba are unclear on one point that I wish to clarify.
Suppose a trait evolves under a particular selective regime because of some particular use. Suppose that later in the history of that lineage the trait, no longer under
the original selection pressure, is co-opted for another use. It is then an exaptation.
But suppose that because of this other use the trait is now under stabilizing selection and remains so for many generations. The trait has not changed, and part of
the reason it has not changed is the history of stabilizing selection. In my view it
then becomes an adaptation because its presence in the population is, in part, due
to selection on that function. That the selection is stabilizing, rather than directional, should make no difference.

172

superior strategy to catch flying prey (review in Coddington, 1986a). Evidence for its adaptive value included the assertion that it had evolved convergently in two distantly related taxa. A cladistic re-analysis of the problem (Coddington, 1986a,b) indicates that the orb web is an ancient behavioral feature of a large group of spider taxa, that it only evolved once and defines a monophyletic taxon, and that cob web weavers are derived from orb weavers, rather than vice versa. Thus, the original hypothesis makes no sense because the presumed polarity of the traits was backwards.

Coddington goes on to say that the orb web might still be an adaptation relative to some (unknown) more primitive feature. But the point is that it did not evolve from the cob web and so its presence cannot be explained in terms of its presumed adaptive advantages over the cob web.

To return to our example of intraspecific adaptation, if metal tolerance were the primitive condition and it did not evolve in response to the selection pressures of contaminated soil, then the presence of metal-tolerant types on contaminated soil would not call for an adaptation explanation. But the presence of nontolerant types in the surrounding pasture would indicate a plausible candidate for an adaptation explanation. In fact, as I mentioned earlier, there are a number of lines of evidence that overwhelmingly support the hypothesis that nontolerance is the primitive condition, and so our adaptation explanation of metal tolerance is based on correct trait polarity.

This raises an interesting possibility. Suppose we were to erect a fence, tall enough to prevent gene flow, 2 meters to the noncontaminated side of a mine boundary. Next we kill all of the plants growing on the 2-meter swath of noncontaminated soil, leaving only the tolerant population growing on the contaminated soil. Because of the past gene flow, seeds from this population contain considerable genetic variation in metal tolerance. These seeds fall on the bare noncontaminated soil and selection for nontolerance occurs. After a number of years we return and find that a nontolerant race has evolved on the noncontaminated soil. Nontolerance (which is probably a bad name for the relevant suite of traits that distinguishes these plants from the metal-tolerant plants and that

makes them better adapted than the tolerant types in noncontaminated soil) would then be an adaptation in that population.

Phylogenetic information on trait polarity is, quite obviously,
historical information. One might be skeptical that this sort of information can generally be recovered. But recent advances in cladistic methodology, as well as new sources of data (e.g., data on
nucleotide sequences), give reasons for optimism. Indeed, all five
components of complete adaptation explanations have a historical
dimension, but it is in the area of phylogenetic determination of
trait polarity that we have the most sophisticated methods of re-
covering history.[15]

### 5.2b. Generalization from the Example

I have argued that a complete adaptation explanation has five
components: (1) Evidence that selection has occurred on the relevant traits; (2) An ecological explanation of that selection; (3) An
account of the underlying genetics of the traits (at minimum, evidence that the relevant traits are heritable); (4) Information on
population structure (in particular on patterns of gene flow and
selective heterogeneity); and (5) Phylogenetic information on trait
polarity.[16] I presented the example of heavy-metal tolerance in
plants to illustrate each of these components. That example is perhaps the most complete adaptation explanation extant in evolutionary biology, but even it is not complete. In particular, not
much is known about the underlying genetics of metal tolerance,
and there is good reason to believe that the structure of the selective environment is not as simple as presence or absence of heavy

[15] Coddington's (1988) article mentioned above contains an excellent discussion
of the relevant methods, as does Greene (1986), Mishler (1988), and Donoghue (in
prep.).

[16] One might compare this list of five components of complete adaptation explanations with Mishler's (1988, p. 288) list of the four tests that a hypothesis of adaptation must pass: "(1) The apomorphic [derived] state of some trait must be
shown to function better (in an engineering sense) in solving some environmental
problem than the plesiomorphic [primitive] state does. (2) The trait must be heritable. (3) The trait must be shown to increase the fitness of organisms bearing it.
(4) The apomorphic trait must have arisen when the improved function arose."
Mishler's (1) corresponds to my (5) plus part of (2). His (2), of course, corresponds
to my (3). His (3) relates to my (1) but is flawed in that it is not explicitly historical.
His (4) draws on my (5) and (2).

metals. Nonetheless, it is a good adaptation explanation. But is it a good model for adaptation explanations in general?

If one believes that the process of adaptation, the process that produces "organs of extreme perfection," is simply an iteration of intraspecific organismic selection, that macroadaptation is simply microadaptation writ large, then our example presumably is a good model. This belief has been quite common among postsynthesis evolutionary biologists. Are there reasons to reject this common view? One reason would *not* be that natural selection is but one of many evolutionary processes. Gould (1983) and others have complained that after the evolutionary synthesis, positions about the role of natural selection hardened and that biologists became less pluralistic with respect to alternative evolutionary mechanisms. We can agree with all of this, but whatever the importance of natural selection relative to other evolutionary mechanisms, only natural selection can lead to adaptations. This is true by definition. Other evolutionary mechanisms, such as drift or pleiotropy, may produce traits that increase the relative adaptedness of their possessors; but not adaptations.[17]

On the other hand, if selection occurs at various levels of biological organization, then our example is limited in that it involves only organismic selection. For instance, group selection may produce group adaptations. In the theoretical models of group selection for altruism, altruism is a group adaptation. But a group adaptation explanation would require the same five components we found in our example of metal tolerance, and so, it seems, our example is a good model for adaptation explanations in general.

At the beginning of section 5.2a I pointed out that the order in which I have listed the five components of a complete adaptation explanation is immaterial. Here I should also point out that there is nothing sacrosanct about my list of five components; there are other ways of organizing these categories. This is so because of the complex interrelations among them. Clearly, point (1), that there

[17] It is worth noting that presumably no serious biologists think that other evolutionary mechanisms, such as drift or pleiotropy, can produce complex and intricate traits that appear to be adaptations. Of course there are alternative "theories" that attempt to explain such traits, creationist and neo-Lamarckian theories among them. Dawkins (1987) offers a good discussion of these, which he calls "doomed rivals." It is beyond the scope of this book to refute these alternatives.

is selection, and point (2), an ecological explanation of selection, are related in that (2) presupposes (1). Three less obvious interconnections bear mentioning. To know that selection has occurred on the *relevant* trait, as opposed to some correlated trait, will oftentimes require knowledge from both (2), ecological genetics, and (3), population genetics. Similarly, (5), phylogenetic information on trait polarity, interconnects with (1) and (2) in reconstructing evolutionary histories; one needs to correlate changes in trait value with changes in selective regimes. As pointed out above, it matters whether heavy-metal tolerance arose prior to or after some lineage's exposure to metal-contaminated soils. Only in the latter case would we hypothesize that the trait was an adaptation. Finally, point (4), information on population structure, bears on (1) in ways that I hope were made clear in chapter 2. (Recall the examples showing that whether or not differential reproduction qualifies as natural selection depends in part on the structure of selective environments.) My claim is that points (1)–(5) are necessary components of complete adaptation explanations. But, for the reasons briefly touched on above, it is clear that the same information could be demarcated differently. My hope is that points (1)–(5) perspicuously demarcate the relevant components.

## 5. 3. How-possibly vs. How-actually Explanations

One decidedly atypical feature of our example of the evolution of heavy-metal tolerance is our degree of access to the information necessary for a complete adaptation explanation. Our explanation of the evolution of heavy-metal tolerance is given in terms of processes that are currently observable. We are confident in the backwards extrapolation from these current processes because we are dealing with relatively recent and rapid evolutionary changes. Although the explanation is not an ideally complete adaptation explanation, it comes close and we can be reasonably sure that, in broad outlines at least, it gives us the mechanisms by which metal tolerance *actually* evolved. In the general case I have outlined reasons for skepticism concerning the construction of ideally complete adaptation explanations. (Imagine coming up with a complete adaptation explanation of the mammalian eye.) I am sure some people will be more, and some less, skeptical than I on this matter. But for the moment let us take the skeptical position that complete or even nearly complete adaptation explanations are going to be rare in evolutionary biology. What are the consequences of this skeptical position?

Although I will not try to defend the view, it can be plausibly maintained that the primary cognitive (as opposed to practical or social) aim of scientific theories is scientific explanation, and that scientific explanation results in scientific understanding of the explained phenomena. So, one might argue, if a theory gives few, if any, complete explanations of the phenomena in its domain, then it is (cognitively) worthless. Applying this reasoning to our case, one would conclude that the theory of evolution by natural selection is without cognitive value. This conclusion is, I will argue, seriously flawed.

The sort of understanding we gain when we have a complete or nearly complete adaptation explanation is easy to appreciate. We can understand a phenomenon such as metal tolerance when we have a complete account of how it actually evolved. But another sort of scientific understanding results from an account of how some adaptation *could have* evolved. We can begin to appreciate this by considering a group of examples that bothered Darwin (1859).

The section entitled "Organs of Extreme Perfection and Complication" in Chapter 6 of *On the Origin of Species* ("Difficulties on Theory") begins as follows (p. 189, emphasis added):

> To suppose that the eye, with all its inimitable contrivances for adjusting the focus to different distances, for admitting different amounts of light, and for the correction of spherical and chromatic aberration, *could have been* formed by natural selection, seems, I freely confess, absurd in the highest possible degree.

This is followed by a plausible account, but one that is clearly speculative and incomplete, of how the eye could have arisen from an "optic nerve merely coated with pigment" (p. 187). Then Darwin says (p. 189, emphasis added):

> If it could be demonstrated that any complex organ existed, which *could not possibly have been* formed by numerous, succes-

sive, slight modifications, my theory would absolutely break down. But I can find out no such case.

For Darwin, at least, explanations of how adaptations could have evolved were of vital importance. The challenge of "organs of extreme perfection" was that, on the face of it, it seemed highly implausible that they could evolve by natural selection. Darwin countered the challenge by providing an account of how they could evolve. Such accounts, which I will call *how-possibly explanations*, have a cognitive value that is independent of whether or not they reflect the way the trait actually evolved. Put another way, how-possibly explanations have a cognitive value that is independent of their truth.[18]

Since I am not presenting a formal theory of explanation, it is difficult for me to precisely characterize how-possibly explanations and their relation to genuine explanations (or how-actually explanations). Hempel and Oppenheim (1948), and Hempel (1965) did give a formal theory of explanation. The formal requirements determine what qualifies as a *potential* explanation, and if the explanatory premises of a potential explanation are true, then it is a genuine explanation.[19] What I am calling how-possibly explanations

[18] I have borrowed the term "how-possibly explanations" from William Dray (1957), although my account of them differs from his. Dray discussed how-possibly explanations as purported counterexamples to Hempel's D-N model of explanation. For Dray a question calling for a how-possibly explanation arises when there is some presumption that implies that some actual phenomenon is impossible, and the explanation consists simply of an account that removes the presumption. For example, on seeing tropical vegetation growing on an island in a glacial lake in southern Switzerland, one might ask how that could possibly happen. For Dray, the explanation consists in removing the presupposition that such vegetation could not live there; thus we answer the question by pointing out that the mountains on the north side of the lake block winds from the north so that the winter climate is quite mild. According to the view I will defend, this is incomplete even as a how-possibly explanation. Removing the presupposition that the climate is too cold for the plants does not fully explain the presence of these plants in Switzerland. For instance, one still wants to know how they got there (in the case I have in mind on Lake Maggiore, horticulturists brought them in).

[19] Hempel's formal requirements for explanation are: (1) The explanation must be a valid deductive argument; (2) among the premises essential for the derivation of the conclusion there must be at least one lawlike generalization; and (3) the premises of the explanans (that which does the explaining) must have empirical content. These requirements are necessary conditions for something to be a potential explanation. To these requirements Hempel added the empirical (and hence nonformal) requirement (4): the premises of the explanans must be true. This was a necessary condition for a potential explanation to be a genuine explanation. As I pointed out

are potential explanations, none of whose explanatory premises contradict or conflict with "known facts" (i.e., the thing we believe based on good evidence). More specifically, if we think, as Hempel did, that explanations of particular events involve both statements of initial conditions and generalizations or laws that connect the initial conditions with the event to be explained, then the sort of explanation I have in mind is one based on generalizations or laws we have good reason to believe are true, but whose initial conditions are speculative.

Let us return to adaptation explanations. What are the relevant generalizations? Foremost among them is the PNS. But other generalizations may also come into play in a complete adaptation explanation; for instance, generalizations from population genetics and ecology and the generalizations that underlie cladistic methods for constructing cladograms (relevant to claims about trait polarity).[20] Any proffered adaptation explanation that conflicts with these generalizations will not be acceptable, not even as a how-possibly explanation. For example, an explanation that posits a single generation fixation of a rare allele in a moderately large population will be rejected as a how-possibly explanation since it conflicts with generalizations from population genetics. On the other hand, consider the relevant initial conditions. As pointed out above, each of the five components of a complete adaptation explanation has a historical dimension. That is to say, each requires information about *contingent* events from the past, for example, past selective environments, patterns of gene flow, and phylogenetic patterns. When I discussed the difficulties presented by each of the five components I was pointing to the problems inherent in making claims about the contingent past. In some cases, such as the case of the evolution of heavy-metal tolerance, we can have solid evidence about the relevant past conditions, but in other

---

earlier, this model of explanation, called the D-N model, is no longer accepted by most philosophers of science. In particular, I agree with Jeffrey (1969) and Salmon (1971, 1984, 1989) that explanations are *not* arguments. Nonetheless, in what follows I will talk of explanatory premises, noting here that this is not meant to imply that explanations are valid deductive arguments. One more note for those who care: by calling (3) a formal requirement, I do not mean to say that one can give a satisfactory formal account of empirical content; that was one of the major goals of Hempel and other logical positivists, but their efforts were not successful.

[20] See Stevens (1980) as well as the articles mentioned in note 14 above.

cases we can only speculate about them. In the latter case we can only offer how-possibly explanations.

We can think of evolutionary biology as being divided into two parts: the first takes as its subject the mechanisms of occurrent evolutionary processes, and the second is devoted to the evolutionary history of life on this planet (Antonovics 1987). Of course, these two parts are interconnected in various ways. For instance, the study of evolutionary history draws heavily on what is known about current evolutionary mechanisms, and the study of evolutionary mechanisms is constrained by what is known about actual evolutionary history. What good is a speculative how-possibly explanation? The short answer is: it shows how known evolutionary mechanisms *could* produce known phenomena. We know mammals have eyes with an intricate structure that allows for the detection of many features of their external environment. How *could* this have evolved? That was Darwin's question, and, as we will see, it is also the sort of question that can be profitably investigated today.

Another of the "organs of extreme perfection" that have been of interest to evolutionists are wings. Their present adaptive significance seems obvious—they are used mostly for flight—but *how* could they have evolved? One answer is that winged creatures evolved from wingless creatures because of a single macromutation and subsequent selection for that new mutant. This answer is not acceptable, however, because it conflicts with what we know from genetics and developmental biology about the likelihood of such a mutation. Another possible answer is that winged creatures gradually evolved from wingless ones where each step in the evolutionary process was driven by the aerodynamic advantages of wings or protowings. But the aerodynamic effects of wings are strongly scale-dependent, so we cannot evaluate this answer abstractly but must restrict it to particular a group of organisms of a particular size. In their article, Joel Kingsolver and M.A.R. Koehl (1985) explore this and other proposed explanations of the evolution of wings in insects. Their study is a perfect example of a how-possibly explanation of an adaptation.

If winged insects did not evolve by one or a few macromutations, then they must have evolved gradually from wingless creatures. Although a gradual evolutionary trajectory can result from drift alone, it is highly improbable that a feature as complex as an insect wing could evolve that way. And so we look for a natural-selectionist explanation of wings. As Kingsolver and Koehl put it, "To understand the evolution of a complex structure like the insect wing, we need to identify its possible adaptive value during the transition from wingless to winged insects" (1985, p. 489). In the terminology of the last section, they are looking for possible eco-logical reasons to explain why insects with small protowings were better adapted than wingless insects. They evaluate several aero-dynamic and thermoregulatory hypotheses that others have of-fered as explanations.

There are three aerodynamic hypotheses: (1) that small proto-wings were used for gliding; (2) that they were used for parachut-ing, to slow the rate of fall; and (3) that they were used for attitude stability, to help the insect land right-side-up. By constructing physical models of approximately the right size and shape (based on the fossil record of Paleozoic insects), Kingsolver and Koehl were able to test these hypotheses in a wind tunnel. Their results show that there are no significant aerodynamic advantages for small protowings: "For insects in the size range from which winged insects probably arose, wing lengths more than 30–60% of body length are required before there are any significant aerody-namic effects of wings" (p. 500). Thus they reject the aerodynamic hypotheses. These hypotheses are not acceptable as how-possibly explanations because they conflict with the laws of aerodynamics.

Kingsolver and Koehl also evaluate the thermoregulatory hy-pothesis by means of physical models. Here the models with var-ious-sized wings were placed under a flood lamp and the effects of the wings on body temperature relative to surrounding air tem-perature were measured. In this case they found that even very small protowings had significant thermoregulatory effects that are of benefit to the insects. Thus the thermoregulatory hypothesis for the evolution of insects with small wings from wingless insects is consistent with the relevant laws or generalizations (which in this case are the laws of thermodynamics).

Of course, this is not the full story. Insect wings are used for flight, and we have good reason to believe that this has played a role in their evolution. Kingsolver and Koehl show that for any body size there is a transition point for relative wing length above which there are no additional thermal effects and below which there are no significant aerodynamic effects; that is, for a wing used as a thermoregulatory device there is a size whose value is determined by body size, such that any increase beyond it will be of no thermoregulatory benefit. But beyond that transition point, thermoregulatory "wings" can become aerodynamic wings. Furthermore, and this is one of their key points, they show that the relative wing length at which this transition occurs decreases with body size. That means that purely isometric increases in body size (for whatever reason) can result in insects whose thermoregulatory devices are "suddenly" aerodynamically useful. Thus they present a two-part scenario for the evolution of insect wings. First came the gradual evolution of relatively small protowings driven by the adaptive advantages resulting from their use as thermoregulatory devices, but this is only one step in the evolution of wings. Next, some insects with well-adapted thermoregulatory devices increased isometrically in size. (Again we can imagine many reasons why there might be selection for increased body size in the relevant population, but that is of no real concern here.) Their wings are suddenly aerodynamically effective, and from here selection for increased aerodynamic efficiency can lead to the present-day insect wings.

Is this the way insect wings *actually* evolved? This we cannot know with certainty. Some of the conditions postulated in this scenario are speculative, and given that we are dealing with the distant past, we cannot reasonably hope that their status will change. But since we can assume that the laws of thermodynamics and aerodynamics have held throughout the evolution of insect wings, certain consequences of the various hypotheses can be tested. These tests led Kingsolver and Koehl to reject the pure aerodynamic hypotheses (as well as the pure thermoregulatory hypothesis) and to formulate the two-part scenario for the evolution of insect wings outlined above. That hypothesis is consistent with all of their experimental results. After commenting on the speculative nature of parts of their scenario, they say, "At best, we can eliminate certain hypotheses as untenable and document other hypotheses as at least plausible" (p. 500).

I have discussed Kingsolver and Koehl's study in some detail for two reasons. First, their study nicely illustrates the value of how-possibly explanations of adaptations. We learn something about the evolutionary process from the scenario Kingsolver and Koehl present. We see how a simple change in body size can result in a change in the function or use of some trait, and how that can result in an evolutionary trajectory that would be impossible without this change in function.[21] Earlier I had mentioned that evolutionary biology can be thought of as divided into two parts—the study of evolutionary mechanisms and the study of evolutionary history. Kingsolver and Koehl's scenario tells us something about the mechanisms of evolution, and it is also a plausible account of the actual evolutionary history. My second reason for discussing their work in some detail is that it counters what I anticipate as a common reaction to my defense of how-possibly explanations of adaptations. In the aftermath of Gould and Lewontin's well-known "Spandrels of San Marco" paper (1979) I am sure some will view how-possibly explanations of adaptations as a sort of "just so" story that Gould and Lewontin rightly vilified. But as we have seen from the above discussion, how-possibly explanations can be rigorously formulated so that they do have testable consequences. Whether the scenario of Kingsolver and Koehl is fundamentally correct or not, no one can fairly describe it as merely an imaginative bit of storytelling.

In the last section we discussed the evolution of heavy-metal tolerance as an example that comes close to being an ideally complete adaptation explanation. We can be confident that, in broad outlines at least, it reflects how that adaptation *actually* evolved. We might call it a *how-actually explanation*. In this section we used Kingsolver and Koehl's scenario for the evolution of insect wings as an example of a how-possibly explanation. Although it is sketchier in many ways than our account of the evolution of metal tolerance, it too may be essentially correct, that is, it too may reflect how the adaptation actually evolved. Thus the distinction between the how-possibly explanation and one we take to be a how-actually or genuine explanation is epistemological.[22] As such it is

[21] Gould (1985) discusses this and other cases of functional shifts. He traces the principle involved back to Darwin.
[22] The epistemological distinction is between how-possibly explanations and *what*

183

not a dichotomous distinction, rather the two examples lie along a continuum. A how-possibly explanation is one where one or more of the explanatory conditions are speculatively postulated. But if we gather more and more evidence for the postulated conditions, we can move the how-possibly explanation along the continuum until finally we count it as a how-actually explanation.

I could say much more in defense of how-possibly explanations. For instance, I view much of the theoretical work in mathematical population genetics as the construction and testing of how-possibly explanations (consider kin and group selection models for the evolution of altruism). But I have said enough to address the question with which we began this section. Suppose it is true that few, if any, ideally complete adaptation explanations will be produced by evolutionary biologists. Does that mean that the theory of evolution by natural selection is cognitively worthless? I hope I have convinced the reader by now that the answer is no.[23]

## 5.4. Mechanistic Explanations of Teleological Phenomena

Section 5.2 was devoted to a detailed explication of the content of ideally complete adaptation explanations, and in the previous section I defended the cognitive value of adaptation explanations that fall short of this ideal in certain ways. It is now time to return to the distinction from the general theory of explanation introduced in section 5.1. This is the distinction between causal/me-

---

*we take to be* a genuine or how-actually explanation. One might still want, à la Hempel, an objective or nonepistemologically relativized account of explanation. I have not presented a formal theory of explanation, but the five conditions set out in section 5.2 can be taken as an informal theory of adaptation explanations. Thus something is a genuine adaptation explanation if it truly meets those conditions, *whether we know it or not.* But my characterization of how-possibly explanations is essentially epistemological. It is based on our knowledge, or lack thereof, of certain explanatory conditions. Thus an objectively genuine, ideally complete adaptation explanation might still be a how-possibly explanation.

[23] Michael Ghiselin (personal communication) has pointed out that the value of how-possibly explanations in evolutionary biology is dependent on the stage of historical development of the field. Thus Darwin's how-possibly explanations of "organs of extreme perfection" were designed to counter impossibility arguments. This was discussed earlier in this section. Later in the development of the field (or presumably within any subfield of evolutionary biology) groups of how-possibly explanations can be used to construct how-actually evolutionary reconstructions.

chanical explanations and unification explanations, or in Kitcher's (1985) terminology, the distinction between bottom-up and top-down explanations.

A causal/mechanical explanation is one that explains the phenomenon of interest in terms of the mechanisms that produced the phenomenon. What is a mechanism? A paradigm example would be a mechanical watch, that is, a spring-powered watch that works through a complicated set of gears. But a contemporary electronic watch is no less a mechanism. Indeed, I have argued (Brandon 1985c) that this question has no general metaphysical answer, because the business of science is the discovery of mechanisms; so we cannot delimit in any a priori manner the mechanisms of nature. Consider, for example, the question of what are the mechanisms of evolutionary change. Clearly this calls for an empirical answer, one that is always subject to revision in light of new discoveries. What is a mechanism? The best we can do is to give an open-ended answer: a mechanism is any describable causal process.

Adaptation explanations are causal/mechanical explanations. We explain an adaptation, such as metal tolerance in plants or wings in insects, in terms of the historical causal process that produced them. Each of the five components of a complete adaptation explanation deals with some aspect of this process. My claim, based on our current knowledge of evolutionary mechanisms, is that these are the five aspects of the process we need to delineate in order to have a complete understanding of it. Because everything that has been said so far about adaptation explanations supports the view that they are causal/mechanical explanations, I will not belabor the point.

Less obvious is the fact that, in one way at least, adaptation explanations fit the unification model of explanation as well. A unification, or top-down, explanation explains phenomena by subsuming them under some central generalization. All adaptation explanations, no matter how diverse the adaptations, subsume the explained adaptation under the PNS (recall the discussion of systematic unification in section 4.2a). Metal-tolerant plants tend to outreproduce nontolerant plants on metal-contaminated soil because they are better adapted to that environment. In the early evolution of insect wings, insects with small protowings tended to outreproduce their conspecifics without wings because they were better adapted, (in that they could better regulate their body temperature). Although adaptedness supervenes on the various organic features, and on their ecological consequences, involved in natural selection (chapter 1), all selection events are subsumed under the PNS. In this way my explication of adaptation explanations fits the unification model of explanation.

Thus with respect to adaptation explanations, the causal/mechanical and unification models of explanations are not in conflict. Rather, a complete adaptation explanation has both unification and causal/mechanical aspects.

But is this the full story on adaptation explanations? Don't birds have wings in order to fly, and bears have eyes in order to see? Isn't it the case that the function of metal tolerance in grasses is to enable them to live on contaminated soil? And what about sex? Doesn't it make sense to question the function of sex? Is it a mechanism to insure genetic variation and hence evolutionary plasticity? Or is it, as some would have it (Bernstein et al. 1984), a mechanism for DNA repair? All of these questions point to the teleological aspect of adaptations. Adaptations seem to be *for* something, they seem to have some special consequences, called *functions*, that seem to help explain their existence. Are adaptation explanations teleological?

In chapter 1 I argued for what I called the historical conception of adaptation, according to which an adaptation is the direct product of the process of evolution by natural selection. This view (or some minor variant of it) has been supported by many authors, including G. C. Williams (1966), Lewontin (1978), Brandon (1978, 1981a), Gould and Vrba (1982), Burian (1983), and Sober (1984). I think it is accurate to call it the received view. An essential part of an adaptation explanation is what I called the ecological explanation of selection. Here we cite the ecological consequences of the adaptation, or its precursors, that explain its adaptive advantage over its alternatives. According to the scenario of Kingsolver and Koehl, insects with protowings were better adapted than their conspecifics lacking them because they were better able to regulate their body temperature. In the case of metal tolerance, some grasses are better adapted than others in an environment of metal-contaminated soils because they are more tolerant of the toxic heavy metals. These ecological consequences that explain evolution by natural selection may be identified as the function of the adaptation. Thus in evolutionary biology, to make a claim about the function of some item is to make a particular claim about its evolutionary history.[24] And, I will argue, to cite the function of an adaptation is to provide a satisfactory answer to a teleological question.

The function of some adaptations is obvious. When an adaptation is described as heavy-metal tolerance, we immediately know its function. But were that very same adaptation differently described, say we described it in terms of its underlying physiology, its function might not be clear. In other cases an apparently obvious function of an adaptation may not in fact be its function. For example, if we found a fossilized insect with a wing 20% of its body length, we might surmise that the wing was for flight. Of course, that guess is contrary to the laws of aerodynamics and would be wrong. In still other cases we may be fairly confident that something is an adaptation but we have no clear idea about its function. G. C. Williams (1966) cites the example of the lateral line in fish. Because of its structural complexity and taxonomic generality, it seemed likely that the lateral line was an adaptation even though its function was unknown. Studies eventually showed that the lateral line was an auditory sense organ. Presumably that is its function, though we should bear in mind the possibility that it is an exaptation for audition and evolved for some

---

[24] I am well aware that this does not accurately describe the usage of the term "function" in all areas of biology. For instance, in physiological or biomechanical studies functional ascriptions usually amount to no more than the identification of some effect of an item that, in some way or other, contributes to the overall working of the organic system. Thus it is said that the function of the heart is to circulate blood. This example is often cited by defenders of a nonevolutionary notion of function, since Harvey discovered the function of the heart two hundred years before Darwin. But the heart has other effects, in particular it produces heart sounds. In modern medicine the sound of a beating heart is a useful diagnostic aid, and so heart sounds also contribute to the overall working of some human organic systems. Is the production of heart sounds a function of the heart? I believe that ahistorical functional ascriptions only invite confusion, and that biologists ought to restrict the concept of its evolutionary meaning, but I will not offer further arguments for that here.

other reason. Whenever we hypothesize that some trait is an adaptation, it makes sense to inquire about its function. I will call this sort of question a *what-for question*. A complete adaptation explanation, in particular the component that gives the ecological explanation of the relevant course of selection, answers this question.

What-for questions are apparently teleological. They can be asked of "small" adaptations, for example, some particular physiological condition in some subpopulation of a species of grass growing along the side of a mine; of "big" adaptations, for example, the lateral line in fish or the mammalian eye; and of even "bigger" adaptations, for example, sexual reproduction. Put abstractly, a what-for question asked of adaptation $A$ is answered by citing the effects of past instances of $A$ (or precursors of $A$) and showing how these effects increased the relative adaptedness of $A$'s possessors (or possessors of $A$'s precursors) and so led to the evolution of $A$. These effects are the function of $A$. The sense in which what-for questions and their answers are teleological can now be clarified. Put cryptically, we explain $A$'s existence in terms of $A$'s function. More fully, $A$'s existence is explained in terms of effects of past instances of $A$; but not just any effects: we cite only those effects relevant to the adaptedness of possessors of $A$.[25] More fully still, adaptation $A$'s existence is explained by a complete adaptation explanation that includes not only the ecological account of the function of the adaptation, but also the other four components detailed in section 5.2.

Adaptations in nature seem to call for teleological explanations. Adaptation explanations, according to my explication, are teleological in the sense that they are answers to what-for questions. But they are also perfectly good causal/mechanical explanations.

---

[25] In Brandon (1981a) I argue that adaptation explanations are significantly different from other evolutionary explanations, either within biology (e.g., an evolutionary explanation given in terms of drift) or in other areas of science (e.g., a cosmological explanation of the evolution of our solar system). The difference has to do with the teleological aspect of adaptation explanations. It reflects differences in the relevant causal processes. The difference is this: in evolution by natural selection certain effects of the traits under selection feed forward in a way that influences their evolutionary success. I argue that this is not the case for other sorts of evolutionary processes. For example, a trait evolving—that is, increasing in relative frequency—in a population by drift may have many effects, but these effects have no bearing on its evolutionary trajectory.

They do not appeal to backwards causation (the effect of the adaptation causing the adaptation to come into existence) or to any sort of all-encompassing, supraempirical design or plan for the biological world. I have entitled this section "mechanistic explanations of teleological phenomena" precisely because: (1) adaptation explanations are thoroughly mechanistic; but (2) they serve to answer teleological questions. Thus the sort of teleology that survives in contemporary evolutionary biology is not only compatible with a mechanistic world view; I have explicated the sense in which what-for questions and their answers are teleological in purely mechanistic terms.[26]

## 5.5. THE LEVELS OF ADAPTATION[27]

In a sense, adaptations are explained in terms of the good they confer on some entities. But what kind of entity is benefited by an adaptation? If we say that an adaptation is for the good of something, what is that something? Richard Dawkins (1982a,b) addresses this question. His answer is, I think, surprising, and if correct it should occasion a radical reorganization of our conception of nature. But in this section I will argue that his answer is not correct. In so doing I hope to shed further light on the nature and value of adaptation explanations.

In chapter 3 we discussed the interactor/replicator distinction. An interactor is an entity that directly interacts with its environment in a way that leads to differential reproduction. Recall the screening-off argument. I argued that interactors always screen off replicators with respect to reproductive success. It follows that interactors are the entities directly exposed to selection. In contrast, a replicator is an entity that directly replicates its structure.

Dawkins (1982b) offers a fourfold classification of replicators: they may be active or passive, and (cutting across that classifica-

[26] Larry Wright (1976) gives a very general account of teleological explanations that seems to cover my account of the teleological aspect of adaptations explanations quite well. But I strongly disagree with one of the basic assumptions of his work, that one can give a mechanism-free account of teleology (see especially p. 97). I reject this approach. In my view the only legitimate way to analyze teleological explanations or the teleology of the explained phenomena is in terms of the underlying mechanisms.

[27] Much of the material for this section comes from Brandon (1985a).

tion) they may be germ-line or dead-end. An active replicator is one that influences its probability of being replicated. For example, any DNA molecule which, either through protein synthesis or the regulation of protein synthesis, has a phenotypic effect, which affects fitness, is an active replicator. A passive replicator is one that does not influence the probability of its being copied. For instance, a gene that has no significant phenotypic effects but is tightly linked to another gene that is undergoing strong selection will increase, or decrease, in frequency because of that linkage, not because of its phenotypic effects. This gene is sometimes called a "hitchhiker" and is a passive replicator.

Replicators, whether active or passive, are classified as germ-line or dead-end depending on whether they are potential ancestors of an indefinitely long line of descendant replicators. A gene in a gamete or, as the name implies, in a germ-line cell in a body is a germ-line replicator. Most of the genes in our bodies are not germ-line and can replicate only a finite number of times, through mitosis.[28] These are dead-end replicators. Note that it is the potential, not the fact, of being in an indefinitely long ancestry that matters for this classificatory distinction. So a gene in a spermatozoon that fails to fertilize an egg is still a germ-line replicator.

This fourfold distinction is necessary for Dawkins's answer to our question concerning the sort of entity benefited by an adaptation. He argues that they are active germ-line replicators (1982b, p. 84):

> The reason active germ-line replicators are important units is that, wherever in the universe they may be found, they are likely to become the basis for natural selection and hence evolution. If replicators exist that are active, variants of them with certain phenotypic effects tend to out-reproduce those with other phenotypic effects. If they are also germ-line replicators, these changes in relative frequency can have long-term, evolutionary impact. The world tends automatically to become

[28] I assume the reader is a member of *Homo sapiens* or some other "higher" animal species. As Buss (1983 and 1987) points out, if you were a plant (or protist or fungus or some "lower" animal) things would be different. All plant cells, and so all plant genes, are potential ancestors of an indefinitely long descendant lineage. Thus all plant genes are germ line, which is to say that the germ-line/dead-end distinction does not hold here. See discussion in section 3.2.

populated by germ-line replicators whose active phenotypic effects are such as to ensure their successful replication. It is these phenotypic effects that we see as adaptations to survival. When we ask *whose* survival they are adapted to ensure, the fundamental answer has to be not the group, nor the individual organism, but the relevant replicators themselves.

Before we evaluate Dawkins's answer, we should first consider whether or not the question is worth answering, that is, is it a legitimate and important question or is it based on inappropriate teleological thinking. Given our analysis of adaptation explanations, can it in any way legitimize saying that adaptations are "for the good of something"? I think it can. Part of a complete adaptation explanation—the ecological explanation of selection—involves showing how the adaptation (or its precursors) increased the relative adaptedness of its possessors (or possessors of its precursors) over their conspecifics that lack the adaptation (or its precursor). Some traits have effects on the adaptedness of their possessors; those traits that increase the relative adaptedness of their possessors and evolve because of that are adaptations. So, with selection as the arbiter of the good, we can say that adaptations are for the good of their possessors, and this locution serves an explanatory function.

But what sorts of entities are the possessors of adaptations? To think, for the moment, in terms of organismic selection, we might naturally think of organisms as the possessors of organismic adaptations. And so we would say generally that an interactor is the kind of entity benefited by adaptations. But couldn't we also say that the genes causally responsible for the ontogenetic production of the adaptation are the possessors of the organismic adaptation, and that they are ultimately the entities benefited by it? If so, then we would say, in agreement with Dawkins, that active germ-line replicators are the ultimate possessors and beneficiaries of adaptations.

Can this dispute be resolved? I believe it can if we attend to the explanatory import of talk of adaptations as being "for the good of something." In chapter 3 (section 3.2) we saw that interactors screen off replicators from differential reproductive success. In cases of organismic selection, the organism's phenotype screens off its genotype with respect to its reproductive success. More-over, this result is perfectly general; it applies at any level of selection. This should not be surprising since proximate causes screen off more remote causes and interactors are, by definition, proximate causes of reproductive success. Recall that this does not deny that replicators are causally relevant to reproductive success, but in this context they are remotely so. It follows that an ecological explanation of selection (the second component of complete adaptation explanations) must be given in terms of differences in properties of interactors. This point is so obvious that I do not see how it could be disputed. (Try to explain why insects with small protowings outreproduce their wingless competitors in an environment where ambient temperatures fluctuate above and below ideal body temperature.) Thus if the locution "adaptation $A$ is for the good of $x$" is to have any explanatory importance, the values of $x$ must always be interactors, the entities that are directly exposed to natural selection—the entities in terms of which explanations of natural selection must be given. And so we can say adaptations are good for interactors.

But isn't it also true that active germ-line replicators benefit from adaptations? This is so, but from the point of view of explaining natural selection and ultimately explaining the evolution of adaptations, it is a truth that has no explanatory value. This can best be appreciated by considering various levels of selection. As we saw in chapter 3, different levels of selection can have the same replicators. For instance, genes, or lengths of DNA, are the replicators in the intracellular selection process that results in the spread of "selfish DNA" (sensu Doolittle and Sapienza 1980, and Orgel and Crick 1980). The same for cases of meiotic drive. In cases of gametic selection it is again the genes, not the whole gamete, that are the replicators, as is also true in cases of selection among sexual organisms and in intrademic group selection (see section 3.4). Although the replicators are the same in all these selection processes, the processes themselves are distinguishable and can result in quite different sorts of adaptations. For example, we would want to distinguish the adaptation of a chromosome that allows it to be disproportionately represented in gametes from adaptations of gamete morphology and from organismic adaptations such as an eye or wing. Finally, we should not confuse group adaptations, such as within-group altruism, with these lower-level adaptations. When we say that adaptations are for the good of the relevant interactors, we draw such distinctions; when we say they are for the good of the relevant replicators, we conflate them into one type of adaptation.

I want to reemphasize that I am not denying the truth of Dawkins's claim that active germ-line replicators benefit from adaptations. Rather, I am saying that in the context of explaining adaptations such a truth has no explanatory value. On the other hand, to say that adaptations are of benefit to the relevant interactors is explanatory and helps us distinguish different levels of adaptation that result from different levels of selection.

If there were only one level of biological organization at which selection occurs, the difference between Dawkins's view and mine would not be significant. Equally good arguments could be mounted for both views, and in the end it would not matter which view we adopted. But it is becoming increasingly evident that we do not live in such a simple world. Theoretical work has shown beyond any reasonable doubt that levels of selection other than the organismic are *possible*, and empirical work has demonstrated the existence of some of these levels of selection. (This is especially true of suborganismal levels, for example, gametic, chromosomal, or genic; see discussion in chapter 3.) Different levels of selection imply that there can be different levels of adaptation. It must be admitted that the most impressive adaptations of nature, the ones that are most intricate and complex, are organismal. This may reflect the fact that organismic selection has been by far the most important level of selection. But it might also reflect a biased search. If we adopt the view that selection occurs *only* at the organismic level, we are in effect putting on a set of theoretical blinders that prevents us from seeing other levels of adaptation. The position that the organismic is the most important level of selection may well be true, but we ought to treat it as an empirical proposition, one in need of testing, rather than as an a priori truth. Thus our concept of adaptation should allow us to draw the relevant distinctions.

The traditional view that there is only one level of selection is without empirical support. It is also contradicted by an impressive body of theoretical work and a small, but growing, body of empirical work. It should be abandoned. Is this the view defended by Dawkins and other genetic selectionists? It seems not to be. Dawkins does admit that there are various levels of what he calls "vehicle" selection, that is, he admits that there are different levels of interactors. (Indeed as I pointed out in chapter 3, he was one of the people who helped draw the interactor/replicator distinction.) But he has a single-level theory of adaptation, and so, as I have argued, his theoretical framework lacks the conceptual tools to explain and differentiate different levels of selection. It is an explanatorily inadequate framework. It too should be abandoned.

## 5.6. EPILOGUE

This chapter has explored the nature and value of adaptation explanations in evolutionary biology. It is, I think, a fitting end to a book with the following two overarching themes: (1) The theory of natural selection needs to be hierarchically expanded so that it has the potential to apply to various levels of biological organization; and (2) natural selection is an ecological process; to understand it we need a conception of the environments in which it occurs. My arguments have at times been long and intricate, but having presented them I can say that both of these themes flow from an in-depth consideration of what is required for explanations of adaptations. Thus I have argued that an explanatorily adequate theory of adaptation requires a hierarchical and ecological theory of natural selection. In this work I have tried to offer the conceptual foundations of such a theory.

Antonovics, J. 1987. The evolutionary dys-synthesis: Which bottles for which wine? *American Naturalist* 129: 321–331.

Antonovics, J.; Bradshaw, A. D.; and Turner, R. G. 1971. Heavy metal tolerance in plants. *Advances in Ecological Research* 7: 1–85.

Antonovics, J.; Clay, K.; and Schmitt, J. 1987. The measurement of small-scale environmental heterogeneity using clonal transplants of *Anthoxanthum ororatum* and *Danthonia spicata*. *Oecologia* 71: 601–607.

Antonovics, J.; Ellstrand, N. C.; and Brandon, R. N. 1988. Genetic variation and environmental variation: expectations and experiments. In *Plant Evolutionary Biology*, ed. by L. D. Gottlieb and S. K. Jain, pp. 275–303. London: Chapman & Hall.

Arnold, A. J., and Fristrup, K. 1982. The theory of evolution by natural selection: A hierarchical expansion. *Paleobiology* 8: 113–129.

Beatty, J. 1981. What's wrong with the received view of evolutionary theory? In *PSA 1980*, vol. 2, ed. by P. Asquith and R. Giere, pp. 397–426. East Lansing, Mich.: Philosophy of Science Association.

———. 1984. Chance and natural selection. *Philosophy of Science* 51: 183–211.

Beatty, J., and Finsen, S. 1989. Rethinking the propensity interpretation. In *What the Philosophy of Biology Is: Essays for David Hull*, ed. by M. Ruse, pp. 17–31. Dordrecht, Holland: Kluwer.

Bell, G. 1982. *The Masterpiece of Nature: The Evolution and Genetics of Sexuality*. Berkeley: University of California Press.

Bernstein, H.; Byerly, H. C.; Hopf, F. A.; Michod, R. A.; and Vemulapalli, G. K. 1983. The Darwinian dynamic. *The Quarterly Review of Biology* 58: 185–207.

Bernstein, H.; Byerly, H. C.; Hopf, F. A.; and Michod, R. A. 1984. Origin of sex. *Journal of Theoretical Biology* 110: 323–351.

Bethell, T. 1976. Darwin's mistake. *Harper's* (February), pp. 70–75.

Bock, W. J. 1980. The definition and recognition of biological adaptation. *American Zoologist* 20: 217–227.

Bock, W. J., and von Wahlert, G. 1965. Adaptation and the form-function complex. *Evolution* 19: 269–299.

Bonner, J. T. 1974. *On Development: The Biology of Form*. Cambridge, Mass.: Harvard University Press.

———. 1980. *The Evolution of Culture in Animals*. Princeton, N.J.: Princeton University Press.

Boyd, R. 1982. Density-dependent mortality and the evolution of social interactions. *Animal Behavior* 30: 972–982.

Boyd, R., and Richerson, P. J. 1985. *Culture and the Evolutionary Process*. Chicago: University of Chicago Press.

Bradshaw, A. D. 1965. Evolutionary significance of phenotypic plasticity in plants. *Advances in Genetics* 13: 115–155.

Brandon, R. N. 1978. Adaptation and evolutionary theory. *Studies in History and Philosophy of Science* 9: 181–206.

———. 1981a. Biological teleology: Questions and explanations. *Studies in History and Philosophy of Science* 12: 91–105.

———. 1981b. A structural description of evolutionary theory. In *PSA 1980*, vol. 2, ed. by P. Asquith and R. Giere, pp. 427–439. East Lansing, Mich.: Philosophy of Science Association.

———. 1982. The levels of selection. In *PSA 1982*, vol. 1, ed. by P. Asquith and T. Nickles, pp. 315–323. East Lansing, Mich.: Philosophy of Science Association.

———. 1985a. Adaptation explanations: Are adaptations for the good of replicators or interactors? In *Evolution at a Crossroads: The New Biology and the New Philosophy of Science*, ed. by B. Weber and D. Depew. Cambridge, Mass.: The MIT Press/A Bradford Book.

———. 1985b. Phenotypic plasticity, cultural transmission, and human sociobiology. In *Sociobiology and Epistemology*, ed. by J. Fetzer. Dordrecht, Holland: Reidel.

———. 1985c. Grene on mechanism and reductionism: More than just a side issue. In *PSA 1984*, vol. 2, ed. by P. Asquith and P. Kitcher, pp. 345–353. East Lansing, Mich.: Philosophy of Science Association.

Brandon, R. N., and Beatty, J. 1984. The propensity interpretation of 'fitness'—no interpretation is no substitute. *Philosophy of Science* 51: 342–347.

Brandon, R. N., and Burian, R. M., eds. 1984. *Genes, Organisms, Populations: Controversies over the Units of Selection*. Cambridge, Mass.: The MIT Press/A Bradford Book.

Brandon, R. N., and Hornstein, N. 1986. From icons to symbols: Some speculations on the origins of language. *Biology and Philosophy* 1: 169–189.

Buchholtz, J. T. 1922. Developmental selection in vascular plants. *The Botanical Gazette* 73: 249–286.

Burian, R. 1983. Adaptation. In *Dimensions of Darwinism*, ed. by M. Grene, pp. 287–314. Cambridge, Mass.: Cambridge University Press.

Buss, L. W. 1983. Evolution, development, and the units of selection. *Proceedings of the National Academy of Science USA* 80: 1387–1391.

———. 1987. *The Evolution of Individuality*. Princeton: Princeton University Press.

Byerly, H. C., and Michod, R. E. 1990. Fitness and evolutionary explanation. *Biology and Philosophy* 4.

Cartmill, M. 1974. *Daubentonia, Dactylopsila*, woodpeckers and klinorhynchy. In *Prosimian Biology*, ed. by R. D. Martin, G. A. Doyle, and A. C. Walker, pp. 656–670. London: Duckworth.

Charnov, E. L., and Krebs, J. R. 1975. The evolution of alarm calls: Altruism or manipulation? *American Naturalist* 109: 107–112.

Clausen, J. and Hiesey, W. M. 1958. Experimental studies on the nature of species. IV. Genetic structure of ecological races. Carnegie Institution of Washington Publication 615.

Clay, K., and Antonovics, J. 1985. Quantitative variation of progeny from chasmogamous and cleistogamous flowers in the grass *Danthonia spicata*. *Evolution* 39: 335–348.

Clements, F. E., and Goldsmith, G. W. 1924. *The Phytometer Method in Ecology*. Carnegie Institution of Washington, Publication No. 356.

Coddington, J. A. 1986a. The monophyletic origin of the orb web. In *Spider Webs and Spider Behavior*, ed. by W. A. Shear. Palo Alto, Calif.: Stanford University Press.

———. 1986b. The genera of the spider family Theridiosomatidae. *Smithsonian Contributions to Zoology* 422: 1–96.

———. 1988. Cladistic tests of adaptational hypotheses. *Cladistics* 4: 3–22.

Cooper, J. P. 1959. Selection and population structure in *Lolium*. I. The initial populations. *Heredity* 13: 317–340.

Cooper, W. S. 1984. Expected time to extinction and the concept of fundamental fitness. *Journal of Theoretical Biology* 107: 603–629.

Crow, J. F. 1979. Genes that violate Mendel's rules. *Scientific American* 240: 134–146.

Damuth, J. 1985. Selection among "species": A formulation in terms of natural functional units. *Evolution* 39: 1132–1146.

Damuth, J., and Heisler, I. L. 1988. Alternative formulations of multilevel selection. *Biology and Philosophy* 3: 407–430.

Darwin, C. 1859. *On the Origin of Species*. London: John Murray.

Dawkins, R. 1976. *The Selfish Gene*. Oxford: Oxford University Press.

———. 1978. Replicator selection and the extended phenotype. *Zeitschrift für Tierpsychologie* 47: 61–76.

———. 1982a. Replicators and vehicles. In *Current Problems in Siciobiology*, ed. by King's College Sociobiology Group, pp. 45–64. Cambridge Eng.: Cambridge University Press.

———. 1982b. *The Extended Phenotype*. Oxford: Freeman.

———. 1987. *The Blind Watchmaker*. New York: W. W. Norton & Co.

Donoghue, M. J. In prep. Cladograms and character evolution.

Doolittle, W. F., and Sapienza, C. 1980. Selfish genes, the phenotype paradigm and genome evolution. *Nature* 284: 601–603.

Dray, W. 1957. *Laws and Explanation in History*. London: Oxford University Press.

Earman, J., ed. 1983. *Testing Scientific Theories*. Minnesota Studies in the Philosophy of Science, vol. 10. Minneapolis: University of Minnesota Press.

Eigen, M.; Gardiner, W.; Schuster, P.; and Winkler-Oswatitsch, R. 1981. The origin of genetic information. *Scientific American* (April): 88–118.

Ekbohm, G.; Fagerstrom, T.; and Agren, G. I. 1979. Natural selection for variation in offspring numbers: Comments on a paper by J. H. Gillespie. *American Naturalist* 114: 445–447.

Eldredge, N. 1985. *The Unfinished Synthesis*. Oxford: Oxford University Press.

Eldredge, N., and Cracraft, J. 1980. *Phylogenetic Patterns and Evolutionary Process*. New York: Columbia University Press.

Eldredge, N., and Salthe, S. 1984. Hierarchy and evolution. In *Oxford Surveys of Evolutionary Biology*, ed. by R. Dawkins and M. Ridley, pp. 184–208. Oxford: Oxford University Press.

Endler, J. A. 1986. *Natural Selection in the Wild*. Princeton, N.J.: Princeton University Press.

Falconer, D. S. 1981. *Introduction to Quantitative Genetics*. New York: Springer-Verlag.

Felsenstein, J. 1976. The theoretical population genetics of variable selection and migration. *Annual Review of Genetics* 10: 253–280.

Fisher, R. A. 1930. *The Genetical Theory of Natural Selection*. Oxford: Oxford University Press.

Friedman, M. 1974. Scientific explanation and understanding. *Journal of Philosophy* 71: 5–19.

Galton, F. 1889. *Natural Inheritance*. New York: Macmillan.

Ghiselin, M. T. 1974. *The Economy of Nature and the Evolution of Sex*. Berkeley: University of California Press.

Gifford, F. 1986. Sober's use of unanimity in the units of selection problem. In *PSA 1986*, vol. 1, ed. by A. Fine and P. Machamer, pp. 473–482. East Lansing, Mich.: Philosophy of Science Association.

Gill, D. E., and Halverson, T. G. 1984. Fitness variation among branches within trees. In *Evolutionary Ecology*, ed. by B. Shorrocks, pp. 105–116. Oxford: Blackwell Scientific Publications.

Gill, D. E.; Berven, K. A.; and Mock, B. A. 1983. The environmental component of evolutionary biology. In *Population Biology: Retrospect and Prospect*, ed. by C. E. King and P. S. Dawson, pp. 1–36. New York: Columbia University Press.

Gillespie, J. H. 1973. Polymorphism in random environments. *Theoretical Population Biology* 4: 193–195.

———. 1974. Natural selection for within-generation variance in offspring number. *Genetics* 76: 601–606.

———. 1975. Natural selection for within-generation variance in offspring number. II. Discrete haploid models. *Genetics* 81: 403–413.

———. 1976. A general model to account for enzyme variation in natural populations. II. Characterization of the fitness functions. *American Naturalist* 110: 809–821.

———. 1977. Natural selection for variances in offspring number: A new evolutionary principle. *American Naturalist* 111: 1010–1014.

———. 1978. A general model to account for enzyme variation in natural populations. V. The SAS-CFF model. *Theoretical Population Biology* 14: 1–45.

Glymour, C. 1980. *Theory and Evidence*. Princeton, N.J.: Princeton University Press.

Goodman, N. 1979. *Fact, Fiction, and Forecast*. 3d ed. Indianapolis: Hackett Publishing Co.

Gould, S. J. 1983. The hardening of the modern synthesis. In *Dimensions of Darwinism*, ed. by M. Grene, pp. 71–93. Cambridge Eng.: Cambridge University Press.

———. 1985. Not necessarily a wing. *Natural History* 94 (October): 12–25.

Gould, S. J., and Eldredge, N. 1977. Punctuated equilibria: The tempo and mode of evolution reconsidered. *Paleobiology* 3: 115–151.

Gould, S. J., and Lewontin, R. C. 1979. The spandrels of San Marco and the Panglossian paradigm: A critique of the adaptationist programme. *Proceedings of the Royal Society of London*, B 205: 581–598.

Gould, S. J., and Vrba, E. 1982. Exaptation—a missing term in the science of form. *Paleobiology* 8: 4–15.

Greene, H. W. 1986. Diet and arboreality in the emerald monitor, *Varanus prasinus*, with comments on the study of adaptation. *Fieldiana Zoology* 31: 1–12.

Hacking, I. 1965. *Logic of Statistical Inference*. Cambridge Eng.: Cambridge University Press.

Hamilton, W. D. 1964. The genetical evolution of social behavior, I and II. *Journal of Theoretical Biology* 7: 1–52.

———. 1975. Innate social aptitudes of man: An approach from evolutionary genetics. In *Biosocial Anthropology*, ed. by R. Fox, pp. 133–155. New York: Wiley.

Harper, J. 1977. *Population Biology of Plants*. London: Academic Press.

Hartl, D. L. 1988. *A Primer of Population Genetics*. 2d ed. Sunderland, Mass.: Sinauer Associates, Inc.

Hedrick, P. W. 1986. Genetic polymorphism in heterogeneous environments: A decade later. *Annual Review of Ecology and Systematics* 17: 535–566.

Hedrick, P. W.; Ginevan, M. E.; and Ewing, E. P. 1976. Genetic polymorphism in heterogeneous environments. *Annual Review of Ecology and Systematics* 7: 1–32.

Hempel, C. 1965. *Aspects of Scientific Explanation*. New York: The Free Press.

Hempel, C., and Oppenheim, P. 1948. Studies in the logic of explanation. *Philosophy of Science* 15: 135–175.

Horn, H. S. 1979. Adaptation from the perspective of optimality. In *Topics in Plant Population Biology*, ed. by A. T. Solbrig, S. Jain, G. B. Johnson, and P. H. Raven, pp. 48–61. New York: Columbia University Press.

Hull, D. 1973. *Darwin and His Critics*. Cambridge, Mass.: Harvard University Press.

———. 1974. *Philosophy of Biological Science*. Englewood Cliffs, N.J.: Prentice-Hall.

———. 1980. Individuality and selection. *Annual Review of Ecology and Systematics* 11: 311–332.

———. 1981. Units of evolution: A metaphysical essay. In *The Philosophy of Evolution*, ed. by U. L. Jensen and R. Harre, pp. 23–44. Brighton, Eng.: Harvester Press.

Humphreys, P. 1981. Aleatory explanation. *Synthese* 48: 225–232.

———. 1983. Aleatory explanation expanded. In *PSA 1982*, vol. 2, ed. by P. D. Asquith and T. Nickles, pp. 208–223. East Lansing, Mich.: Philosophy of Science Association.

Jablonski, D. 1986. Background and mass extinctions: The alternation of macroevolutionary regimes. *Science* 231: 129–133.

Jeffrey, R. C. 1969. Statistical explanation vs. statistical inference. In *Essays in Honor of Carl G. Hempel*, ed. by N. Rescher, pp. 104–113. Dordrecht, Holland: D. Reidel Publishing Co.

Kettlewell, H.B.D. 1955. Selection experiments on industrial melanism in the Lepidoptera. *Heredity* 9: 323–342.

———. 1956. Further selection experiments on industrial melanism in the Lepidoptera. *Heredity* 10: 287–301.

———. 1973. *The Evolution of Melanism: The Study of a Recurring Necessity*. Oxford: Oxford University Press.

Kimura, M. 1983. *The Neutral Theory of Molecular Evolution*. Cambridge, Eng.: Cambridge University Press.

King, J. L., and Jukes, T. H. 1969. Non-Darwinian evolution. *Science* 164: 788–798.

Kingsolver, J. G., and Koehl, M.A.R. 1985. Aerodynamics, thermoregulation, and the evolution of insect wings: Differential scaling and evolutionary change. *Evolution* 39: 488–504.

Kitcher, P. 1981. Explanatory unification. *Philosophy of Science* 48: 507–531.

———. 1985. Two approaches to explanation. *Journal of Philosophy* 82: 632–639.

———. 1989. Explanatory unification and the causal structure of the world. In *Scientific Explanation*, ed. by P. Kitcher and W. C. Salmon, Minneapolis: University of Minnesota Press.

Klekowski, E. J. 1976. Mutation load in a fern population growing in a polluted environment. *American Journal of Botany* 63: 1024–1030.

———. 1979. The genetics and reprodutive biology of ferns. In *The Experimental Biology of Ferns*, ed. by A. F. Dyer, pp. 133–169. London: Academic Press.

———. 1982. Genetic load and soft selection in ferns. *Heredity* 49: 191–197.

———. 1984. Mutational load in clonal plants: A study of two fern species. *Evolution* 38: 417–426.

Lack, D. 1954. The evolution of reproductive rates. In *Evolution as a Process*, ed. by J. S. Huxley, A. C. Hardy, and E. B. Ford, pp. 143–156. London: Allen & Unwin.

Lace, E. P.; Real, L.; Antonovics, J.; and Heckel, D. G. 1983. Variance models in the study of life histories. *American Naturalist* 122: 114–131.

Lewontin, R. C. 1968. The concept of evolution. *International Encyclopedia of the Social Sciences*, pp. 202–210. New York: Macmillan.

——. 1970. The units of selection. *Annual Review of Ecology and Systematics* 1: 1–18.

——. 1974. *The Genetic Basis of Evolutionary Change*. New York: Columbia University Press.

——. 1978. Adaptation. *Scientific American* 239 (9): 156–169.

——. 1983. Gene, organism and environment. In *Evolution from Molecules to Men*, ed. by D. S. Bendall, pp. 273–285. Cambridge, Eng.: Cambridge University Press.

Lloyd, E. A. 1988. *The Structure and Confirmation of Evolutionary Theory*. Westport, Conn.: Greenwood Press.

Lovtrup, S. 1975. Letter. *Systematic Zoology* 24: 507–511.

MacArthur, R. H., and Connell, J. H. 1967. *The Biology of Populations*. New York: John Wiley.

Macbeth, N. 1971. *Darwin Retried: An Appeal to Reason*. Boston: Gambit.

McNeilly, T. 1968. Evolution in closely adjacent plant populations. III. *Agrostis tenuis* on a small copper mine. *Heredity, London* 23: 99–108.

Malthus, T. R. 1798. *An Essay on the Principle of Population*. London: Johnson.

Mates, B. 1972. *Elementary Logic*. 2d ed. New York: Oxford University Press.

Matessi, C., and Karlin, S. 1984. On the evolution of altruism by kin selection. *Proceedings of the National Academy of Science USA* 81: 1754–1758.

Maynard Smith, J. 1976. Group selection. *Quarterly Review of Biology* 51: 277–283.

——. 1978. *The Evolution of Sex*. Cambridge, Eng.: Cambridge University Press.

——. 1982. The evolution of social behavior—a classification of models. In *Current Problems in Sociobiology*, ed. by Cambridge Sociobiology Group, pp. 29–44. Cambridge, Eng.: Cambridge University Press.

——. 1987. How to model evolution. In *The Latest on the Best: Essays on Evolution and Optimality*, ed. by J. Dupre, pp. 119–131. Cambridge, Mass.: The MIT Press/Bradford Books.

Mayo, D. G., and Gilinsky, N. L. 1987. Models of group selection. *Philosophy of Science* 54: 515–538.

Mayr, E. 1961. Cause and effect in biology. *Science* 134: 1501–1506.

——. 1963. *Animal Species and Evolution*. Cambridge, Mass.: Harvard University Press.

——. 1978. Evolution. *Scientific American* 239: 46–55.

——. 1982. Adaptation and selection. *Biologisches Zentralblatt* 101: 161–174.

——. 1987. The ontological status of species: Scientific Progress and Philosophical Terminology. *Biology and Philosophy* 2: 145–166.

Michod, R. E. 1982. The theory of kin selection. *Annual Review of Ecology and Systematics* 13: 23–55.

Michod, R. E. 1983. Population biology of the first replicators. *American Zoologist* 23: 5–14.

———. 1984. Genetic constraints on adaptation, with special reference to social behavior. In *The New Ecology: Novel Approaches to Interactive Systems*, ed. by P. W. Price, C. N. Slobodchikoff, and W. S. Gaud, pp. 253–279. New York: Wiley.

———. 1985. On adaptedness and fitness and their role in evolutionary explanation. *Journal of the History of Biology* 19: 289–302.

Michod, R. E., and Levin, B. R., eds. 1988. *The Evolution of Sex*. Sunderland, Mass.: Sinauer Associates.

Mills, S., and Beatty, J. 1979. The propensity interpretation of fitness. *Philosophy of Science* 46: 263–286.

Mishler, B. D. 1988. Reproductive ecology of bryophytes. In *Plant Reproductive Ecology: Patterns and Strategies*, ed. by J. Lovett Doust and L. Lovett Doust, pp. 285–306. Oxford: Oxford University Press.

Mishler, B. D., and Brandon, R. N. 1987. Individuality, pluralism, and the phylogenetic species concept. *Biology and Philosophy* 2: 37–54.

Mishler, B. D., and Donoghue, M. J. 1982. Species concepts: A case for pluralism. *Systematic Zoology* 31: 491–503.

Mulcahy, D. L., ed. 1975. *Gamete Competition in Plants and Animals*. Amsterdam: North Holland Publishing Company.

Munson, R. 1971. Biological adaptation. *Philosophy of Science* 38: 200–215.

Nagel, E. 1977. Teleology revisited. *Journal of Philosophy* 74: 261–301.

Nunney, L. 1985. Group selection, altruism and structured-deme models. *American Naturalist* 126: 212–230.

Orgel, L. E., and Crick, F.H.C. 1980. Selfish DNA: The ultimate parasite. *Nature* 284: 604–607.

Peters, R. 1976. Tautology in evolution and ecology. *American Naturalist* 110: 1–12.

Popper, K. R. 1959. The propensity interpretation of probability. *British Journal of Philosophy* 10: 25–42.

———. 1972. *Objective Knowledge*. Oxford: Oxford University Press.

———. 1974. Intellectual autobiography. In *The Philosophy of Karl Popper*, ed. by P. Schilpp. LaSalle, Ill.: Open Court.

Quine, W.V.O. 1951. Two dogmas of empiricism. *Philosophical Review* 60: 20–43.

———. 1960. *Word and Object*. Cambridge, Mass.: The MIT Press.

Railton, P. 1978. A deductive-nomological model of probabilistic explanation. *Philosophy of Science* 45: 206–226.

———. 1980. Explaining explanation. Ph.D. dissertation, Princeton University.

———. 1981. Probability, explanation, and information. *Synthese* 48: 233–256.

Rausher, M. D. 1984. The evolution of habitat preference in subdivided populations. *Evolution* 38: 596–608.

Rausher, M. D., and Englander, R. 1987. The evolution of habitat preference. II. Evolutionary genetic stability under soft selection. *Theoretical Population Biology* 31: 116–139.

Reichenbach, H. 1949. *The Theory of Probability* 2d ed. Berkeley: University of California Press.

Rosenberg, A. 1978. Supervenience of biological concepts. *Philosophy of Science* 45: 368–386.

———. 1982. On the propensity interpretation of fitness. *Philosophy of Science* 49: 268–273.

———. 1983. Fitness. *Journal of Philosophy* 80: 457–473.

———. 1985. *The Structure of Biological Science*. Cambridge, Eng.: Cambridge University Press.

Roughgarden, J. 1979. *Theory of Population Genetics and Evolutionary Ecology: An Introduction*. New York: Macmillan.

Ruse, M. 1971. Functional statements in biology. *Philosophy of Science* 38: 87–95.

———. 1975. Charles Darwin's theory of evolution: An analysis. *Journal of the History of Biology* 8: 219–241.

Russell, B. 1948. *Human Knowledge: Its Scope and Limits*. London: George Allen & Unwin Ltd.

Salmon, W. C. 1971. *Statistical Explanation and Statistical Relevance*. Pittsburgh: University of Pittsburgh Press.

———. 1984. *Scientific Explanation and the Causal Structure of the World*. Princeton, N.J.: Princeton University Press.

———. 1989. Four decades of scientific explanation. In *Scientific Explanation*, ed. by P. Kitcher and W. C. Salmon. Minneapolis: University of Minnesota Press.

Salthe, S. N. 1985. *Evolving Hierarchical Systems*. New York: Columbia University Press.

Scheffler, I. 1963. *Anatomy of Enquiry*. Indianapolis: Bobbs-Merrill Co.

Scriven, M. 1959. Explanation and prediction in evolutionary theory. *Science* 130: 477–482.

Slatkin, M. 1973. Gene flow and speciation in a cline. *Genetics* 75: 733–756.

———. 1981. Population heritability. *Evolution* 35: 859–871.

Smart, J.J.C. 1963. *Philosophy and Scientific Realism*. London: Routledge and Kegan Paul.

Sober, E. 1984. *The Nature of Selection*. Cambridge, Mass.: The MIT Press/ A Bradford Book.

———. 1987. Comments on Maynard Smith's "How to Model Evolution." In *The Latest on the Best: Essays on Evolution and Optimality*, ed. by J. Dupre, pp. 133–145. Cambridge, Mass.: The MIT Press/Bradford Books.

Sober, E., and Lewontin, R. 1982. Artifact, cause, and genic selection. *Philosophy of Science* 47: 157–180.

Stanley, S. M. 1975. A theory of evolution above the species level. *Proceedings of the National Academy of Sciences USA* 72: 646–650.

Stanley, S. M. 1979. *Macroevolution: Pattern and Process.* San Francisco: W. H. Freeman.

Sterelny, K., and P. Kitcher. 1988. The return of the gene. *Journal of Philosophy* 85: 339–361.

Stern, J. T. 1970. The meaning of "adaptation" and its relation to the phenomenon of natural selection. *Evolutionary Biology* 4: 39–66.

Stevens, P. F. 1980. Evolutionary polarity of character states. *Annual Review of Ecology and Systematics* 11: 333–358.

Sultan, S. E. 1987. Evolutionary implications of phenotypic plasticity in plants. In *Evolutionary Biology*, vol. 21, ed. by M. K. Hecht, B. Wallace, and G. T. Prance, pp. 127–178. New York: Plenum Publishing Corp.

Suppe, F. 1977. *The Structure of Scientific Theories.* 2d ed. Urbana: University of Illinois Press.

Thoday, J. M. 1953. Components of fitness. *Symposium of the Society for Experimental Biology* 7: 96–113.

———. 1958. Natural selection and biological process. In *A Century of Darwin*, ed. by S. A. Barnett, pp. 313–333. London: Heinemann.

Turkington, R., and Harper, J. L. 1979. The growth, distribution and neighbour relationships of *Trifolium repens* in a permanent pasture. IV. Fine scale biotic differentiation. *Journal of Ecology* 67: 245–254.

Turkington, R.; Cahn, M. A.; Vardy, A; and Harper, J. L. 1979. The growth distribution and neighbour relationships of *Trifolium repens* in natural and disturbed sites. *Journal of Ecology* 67: 231–243.

Uyenoyama, M., and Feldman, M. W. 1980. Theories of kin and group selection: A population genetics perspective. *Theoretical Population Biology* 19: 87–123.

Van Valen, L. 1971. Adaptive zones and the orders of mammals. *Evolution* 25: 420–428.

———. 1976. Ecological species, multispecies, and oaks. *Taxon* 25: 233–239.

Via, S., and Lande, R. 1985. Genotype-environment interaction and the evolution of phenotypic plasticity. *Evolution* 39: 505–522.

Vrba, E. S. 1984. Evolutionary pattern and process in the sister-group Alcelaphini-Aepycerotini (Mammalia: Bovidae). In *Living Fossils*, ed. by N. Eldredge and S. M. Stanley, pp. 62–79. New York: Springer-Verlag.

Vrba, E. S., and Gould, S. J. 1986. The hierarchical expansion of sorting and selection: Sorting and selection cannot be equated. *Paleobiology* 12: 217–228.

Wade, M. J. 1976. Group selection among laboratory populations of *Tribolium. Proceedings of the National Academy of Science USA* 73: 6404–6407.

———. 1977. An experimental study of group selection. *Evolution* 31: 134–153.

———. 1978. A critical review of the models of group selection. *Quarterly Review of Biology* 53: 101–114.

———. 1984. Soft selection, hard selection, kin selection, and group selection. *American Naturalist* 125: 61–73.

Wade, M. J., and Breden, F. 1981. The effect of inbreeding on the evolution of altruistic behavior by kin selection. *Evolution* 35: 844–858.

Waters, C. K. 1985. Environments, pragmatics, and genic selectionism. Presented at the 1985 Eastern Division Meeting of the American Philosophical Association.

———. 1986. Natural selection without survival of the fittest. *Biology and Philosophy* 1: 207–225.

Williams, G. C. 1966. *Adaptation and Natural Selection*. Princeton, N.J.: Princeton University Press.

———. 1975. *Sex and Evolution*. Princeton, N.J.: Princeton University Press.

Williams, M. B. 1970. Deducing the consequences of evolution: A mathematical model. *Journal of Theoretical Biology* 29: 343–385.

———. 1981. Similarities and differences between evolutionary theory and the theories of physics. In *PSA 1980*, vol. 2, ed. by P. Asquith and R. Giere, pp. 385–396. East Lansing, Mich.: Philosophy of Science Association.

Wilson, D. S. 1979. Structured demes and trait-group variation. *American Naturalist* 113: 606–610.

———. 1980. *The Natural Selection of Populations and Communities*. Menlo Park, Calif.: Benjamin/Cummings.

———. 1983a. The group selection controversy: History and current status. *Annual Review of Ecology and Systematics* 14: 159–187.

———. 1983b. The effect of population structure on the evolution of mutualism: A field test involving burying beetles and their phoretic mites. *American Naturalist* 121: 851–870.

Wimsatt, W. C. 1980. Reductionistic research strategies and their biases in the units of selection controversy. In *Scientific Discovery* vol. 2, *Historical and Scientific Case Studies*, ed. by T. Nickles, pp. 213–259. Dordrecht, Holland: Reidel.

———. 1981. The units of selection and the structure of the multi-level genome. In *PSA 1980*, vol. 2, ed. by P. Asquith and R. Giere, pp. 122–183. East Lansing, Mich.: Philosophy of Science Association.

Wright, L. 1976. *Teleological Explanations*. Berkeley: University of California Press.

Wright, S. 1946. Isolation by distance under diverse systems of mating. *Genetics* 31: 39–59.

Wynne-Edwards, V. C. 1962. *Animal Dispersion in Relation to Social Behaviour*. Edinburgh and London: Oliver and Boyd.